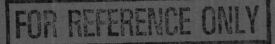

# THE HISTORY OF
# MAN-POWERED FLIGHT

# THE HISTORY OF
# MAN-POWERED FLIGHT

by

## D. A. REAY

*International Research & Development Co. Ltd.,*
*Newcastle-upon-Tyne, England*

## PERGAMON PRESS

OXFORD · NEW YORK · TORONTO · SYDNEY · PARIS · FRANKFURT

| U.K. | Pergamon Press Ltd., Headington Hill Hall, Oxford OX3 0BW, England |
|---|---|
| U.S.A. | Pergamon Press Inc., Maxwell House, Fairview Park, Elmsford, New York 10523, U.S.A. |
| CANADA | Pergamon of Canada Ltd., 75 The East Mall, Toronto, Ontario, Canada |
| AUSTRALIA | Pergamon Press (Aust.) Pty. Ltd., 19a Boundary Street, Rushcutters Bay, N.S.W. 2011, Australia |
| FRANCE | Pergamon Press SARL, 24 rue des Ecoles, 75240 Paris, Cedex 05, France |
| FEDERAL REPUBLIC OF GERMANY | Pergamon Press GmbH, 6242 Kronberg-Taunus, Pferdstrasse 1, Federal Republic of Germany |

First edition 1977

**British Library Cataloguing in Publication Data**
Reay, David Anthony.
Man-powered flight.

Includes index.
1. Human-powered aircraft. I. Title.
TL769.R4 1977        629.133'3        77-7924
ISBN 0-08-021738-9

*In order to make this volume available as economically and as rapidly as possible the author's typescript has been reproduced in its original form. This method unfortunately has its typographical limitations but it is hoped that they in no way distract the reader.*

*Printed in Great Britain by William Clowes & Sons, Limited London, Beccles and Colchester*

# Contents

Contents

# The Author

David Reay, who is 34, is a Chartered
Engineer.  Born in Northumberland, he
was educated at Giggleswick School,
Yorkshire, and Bristol University, where
he graduated in 1965 with a B.Sc. in
Aeronautical Engineering.  His interest
in man-powered flight arose during work
at Hawker Siddeley Aviation, Brough,
and he subsequently spent a further year
at Bristol carrying out research into
the aerodynamics of wings when flying
close to the ground, with particular
reference to man-powered aircraft.

He returned to North East England in
1966, joining C.A. Parsons & Co. Ltd.
as a research aerodynamicist.  Two years
later he moved to the International Research & Development Co. Ltd., also in
Newcastle upon Tyne, where he is Energy
Adviser and is active in the fields of
heat transfer and fluid flow.  He is an
authority on the subject of heat pipe
technology.

Already established as an author of
technical books, these including *Heat
Pipes*, written in conjunction with Professor P.D. Dunn of Reading University,
and *Industrial Energy Conservation*, both
of which are published by Pergamon Press,
he maintains a strong interest in aviation, including man-powered flight and
aircraft photography. He is a member of
the Royal Aeronautical Society Man- Powered Aircraft Group.

His activities extend to include work
on physiology, in particular heat transfer through the skin and the functioning
of sweat glands, a subject not unrelated
to man-powered flight!  In collaboration
with staff of the Department of Dermat-
ology at Nijmegen University in The Neth-
erlands, he has contributed a Chapter to
a major dermatological handbook published
by Springer Verlag, Berlin.  This section
is also being published as a *Monograph on
Sweat Gland Behaviour*.

# Preface

The idea to write a history of man-powered flight occurred at a time when my own interest in the subject was widening through activities associated with the design of a man-powered aircraft within my own company. Books devoted to man-powered flight are few, and this is the first comprehensive history of the topic to be published in the English language. I believe it appears at an appropriate time in the evolution of this form of flight; transition from the ridiculous to the sublime began in the 1920s, and designs have reached maturity during the past fifteen years, although a colourful smattering of the ridiculous still remains.

Serious proponents of man-powered flight are divided as to the direction future developments should take. Some are striving, I believe quite rightly, for a form of perfection epitomised by the requirements of the current major competition, the Kremer Competition, initiated in the United Kingdom, which is spurned by an equally able and enthusiastic faction who see the activity as a sport with, initially, more limited objectives.

Because man-powered flight has an almost universal attraction as a human endeavour yet to be completely fulfilled, the interest in, and appeal of, the subject is not restricted to experts in the field of aeronautics. This book is directed at all those with an interest in aviation, laymen and technical men alike, and, as the development of the science of man-powered flight has progressed until recently largely on an ad hoc experimental basis, I have been able to avoid the use of excessive mathematics in my treatment of the subject. The reader who wants more specialist information will find several complementary books and papers dealing in depth with the theory of flight related to man-powered aircraft in particular, and aircraft in general.

I have been very fortunate in obtaining assistance from many constructors of man-powered aircraft, both past and present, and others associated with equally important but less dramatic aspects of these aircraft. Such a comprehensive history would not have been possible without their co-operation. I am also most grateful to Miss E.C. Pike, former Secretary of the Royal Aeronautical Society Man-Powered Aircraft Group, Mr. K. Clark, current Secretary and to Mr. B.S. Shenstone, founder member of the Cranfield Man-Powered Aircraft Committee, for the loan of photographs and for giving me access to information collected by them over a period of many years.

David A. Reay

August 1976.

# Photo Credits

The copyright of illustrations used in this book belongs to the following agencies:

Radio Times Hulton Picture Library: *Figs. 1, 3, 5, 8, 9, 12, 15, 16, 17, 18, 20, 22, 24, 25, 30, 34, 35.*

Science Museum Photographic Library: *Figs. 2, 4, 6, 7, 11, 13, 19, 26.*

Royal Aeronautical Society: *Figs. 33, 40, 44, 124, 164.*

Evening Standard (London): *Fig. 96.*

Keystone Press Agency: *Figs. 97, 97, 162.*

United Press International: *Figs. 90, 94, 95, 139.*

Flight International: *Figs. 92, 101.*

Associated Newspapers: *Figs. 118, 119, 120, 140, 142.*

Sport & General Press: *Figs. 99, 138.*

Central Press Photos: *Fig. 117.*

South Bedfordshire News Agency: *Fig. 97.*

'Punch' Magazine: *Fig. 145.*

Aireview (Japan): *Figs. 131, 132, 133, 134, 135.*

# Mythology
## or
## Fact
### *Flight in Early Civilisations*

# 1

Man's desire to fly has persisted throughout his history and complete
fulfilment of his wish has only occurred in the twentieth century.  Even today,
romantic associations with flying are still strong, and, as a means of travel,
aircraft have always had an exciting appeal.  It is of course to be expected
that in his earliest attempts at flight man should try to imitate the birds.
Certainly, early man could conceive of no other way of flying than that shown
to him by nature, unless it was to harness birds to lift his body into the air.

## MYTHOLOGY

Early tales of flights of this kind are mainly confined to the realms of
mythology.  Undoubtedly the most popular and widely known mythological flying
escapade is that attributed to Icarus and Daedalus.  Legend has it that
Daedalus, an engineer responsible for the layout and construction of the
labyrinth for the Minotaur in Crete, was imprisoned, together with his son
Icarus, by the ruler of Crete, King Minos.  This Greek story, reputed to date
from about 3500 B.C., goes on to tell how Daedalus constructed two sets of
wings, using mainly wax and feathers as shown in *Fig. 1*.  The father and son
then strapped the wings to their arms, intending to fly from Crete across the
channel separating it from the mainland.  The take-off went well and the flight
was without incident until Icarus, no doubt intoxicated by his new-found
freedom, disregarded his father's warnings and climbed to a considerably
greater altitude.  Flying too close to the sun, the wax supporting the
feathers melted, his wings disintegrated, and the unfortunate Icarus plunged
into the sea and was drowned.  Daedalus flew on and reached the safety of the
mainland, as illustrated in *Fig. 2*.

Other branches of Greek mythology refer to winged creatures with a
basically human form.  Eos, goddess of the Dawn and Artemis both supported
wings.  The personifications of Death, Sleep, Fame, Victory and, appropriately
the Winds, were all shown with similar appendages.  The Greek god Hermes was
winged, as was Cupid.  Phaethon, the son of the Greek sun-god Helios, was
thought to drive a great winged chariot across the sky.  Perseus and Pegasus,
shown in *Fig. 3*, were also depicted in this manner.

Another early legend, in this instance dealing with the flight of a man
aided by a bird, is illustrated on Babylonian seal cylinders dating from the
period 2500 - 2000 B.C.  The legend relates how Etana, the shepherd king was
carried to the aerial palace of the goddess Ishtar by an eagle that he had
befriended.  It tells that the reason for his journey was to intercede on
behalf of his capital city, but his flight was brought to an abrupt conclusion
and he fell to earth.

There are many other references to birds helping men to achieve flight.
Some describe how men soared into the air mounted on the backs of birds,

others refer to the use of chariots or other vehicles hauled by large flocks
of birds. In particular, story tells how King Kai Kawus[1] of Persia harnessed
eagles to his throne. The eagles, illustrated in *Fig. 4*, had been trained and
strengthened for the task since birth. An account of his flights written in
the tenth century refers to the use of the equivalent of a donkey's carrot to
lure the eagles into the air, carrying the king's throne. At each of four
corners of the throne a javelin was stuck into the ground, its point vertical.
Goat's flesh was hung on the top of each spear, and as the eagles became hungry,
so their attempts to reach the food became greater, until enough lift was
generated to raise the king from the ground.

## RELIGIOUS IDENTIFICATION

Flight was not solely the prerogative of man and the birds – winged horses
and bulls have also appeared fairly frequently. Human headed winged bulls[2] of
enormous proportions were mystic guardians of the portals of the palace of
Sargon, situated at Khorsabad, and were emulated by his son Sennacherib at
Ninevah and also by King Xerxes at Persepolis. The Christian religion was not
the only one in which flight was necessary 'to cross the Styx'. Kings and
priests were shown to wear wings when portrayed in Assyrian reliefs during
religious ceremonies. On the sculptured walls of the palace at Nimrud, Ashur-
Nasir-Pal was shown wearing wings during the performance of sacrificial rites,
and scholars believe that the wings were worn on these occasions by those who
were deemed the representatives, and in some cases the incarnation, of the deity.

The Christian religion is to some considerable extent symbolised by winged
beings, no doubt partly because of the early belief that Heaven was in the sky
and flying was the sole form of motion likely to enable travel between Heaven
and Earth. Angels are normally thought of as winged, and many and varied
descriptions are revealed in the Bible. Christian artists have represented
and interpreted angels in many ways. The number of wings depends to a large
extent on the function performed by these spiritual representatives. As
classified by Dionysius, angels were arranged into nine ranks or orders, which
were grouped into three main divisions. Angels, Archangels and Principalities
constituted the 'ministers' whose assigned function it was to communicate
directly with man. The Powers, Virtues and Dominions were 'governors' who
relayed divine commands. The highest order comprised the Thrones, Cherubim
and Seraphim, known as 'councillors'. Only in this last category was a
departure made from the normal representation of two-winged beings with human
form. The Thrones were often portrayed as two fiery wheels with four wings,
Cherubim with six wings and Seraphim, reputed to embody the three virtues of
spirit, mobility and love, also had six wings: "Each had six wings; with two
he covered his face, and with two he covered his feet, with two he flew".

## PORTRAYAL OF FLIGHT BY EARLY CIVILISATIONS

In early Chinese and Indian civilisations, advanced thought touched upon
similar ideas to those of European developed cultures. The mythologies of
Far Eastern civilisations often involved winged creatures, more often demons
than gods.

---

[1] *In the earliest records he is called Kaoos.*

[2] *Blanche Stillson. Wings. Insects, Birds, Men. Gollancz. 1955.*

In the great Indian epic, the *Mahabharata*, reference is made to an aerial war chariot lifted by wings, and Sanskrit literature in general abounds with notes on flying machines. The Indians are attributed with two distinct forms of aerial vehicle; one of these was constructed on the principal of bird flight, and the other was thought to have been derived from a system of flight envisaged by the Greeks. This latter system was kept a closely guarded secret.

About the eighth century, Buddhist 'Apsaras' or angels were depicted in paintings. In parallel with the development of the angels of Christianity were the 'Hsien' or perfected immortals. These were represented as possessing feathered wings, or wearing an outer garment covered with bird feathers.

## EARLY CHINESE DOCUMENTED FLIGHTS

Some of the earliest documented attempts at serious investigation of the possibilities of man-powered flight, and the first claim for a successful flight, originate in China[1]. Emperor Shun, who lived from 2258 B.C. to 2208 B.C. is reputed to have been taught rudimentary aerodynamics by the daughters of Emperor Yao. Apparently he had the opportunity to put theory into practice at first hand. He was constructing a granary, and, finding himself trapped on the roof because the building had caught fire, he "donned the work clothes of a bird, and flying made his escape". Other references quote the use of two large straw hats as parachutes.

Needham, in his treatise entitled *Science and Civilisation in China*, goes into considerable detail when describing early experiments on man-powered flight. He states that self-propelled flying carts were the subject of the earliest Chinese references to flight, as opposed to man using his muscles. The fourth century saw a model helicopter top, to be seen much later in Western civilisation.

An account of the construction of a model bird, made from wood with "wings and pinions, having in its belly a mechanism which enabled it to fly several 'Li'[2]" is given in the book *Wen Shih Chuan*, written by Chang Yin. The design of this flying machine was the work of Chang Hêng, an astronomer and engineer who lived from A.D. 78 to 139. Needham comments that forward propulsion could conceivably have been obtained by using the airscrew development of the helicopter top, but the only power source available at that time would have been a spring, apart from muscle power. (A western parallel to this is the 'Flying Dove' of Archytas of Tarentum, a Pythagorean).

An extremely obnoxious and tyrannical ruler, Emperor Wen Hsiian Ti of the Northern Chhi, who ruled during the short Kao Yang dynasty from A.D. 550 to 551, was noted for a singularly frightening method of punishment; he forced his prisoners to become 'guinea pigs' to further his experiments into man-powered ornithopter flight. One account of his cruel methods recounts how the Emperor visited the Tower of the Golden Phoenix, this being approximately 30 metres high, to receive Buddhist ordination. He ordered many prisoners condemned to death to be brought forward, and instructed that they be harnessed with large

---

[1] J. Needham. *'Science and Civilisation in China'* Vol. IV, *Physics and Physical Technology. Cambridge University Press, 1965.*

[2] A *'Li'* is approximately 0.5 km.

bamboo mats as wings.  They were then ordered to jump from the top of the tower, gliding or flapping their wings.  Emperor Wên Hsiian justified his actions in what was of course solely a form of execution by a convenient mis-interpretation of the practice of 'the liberation of living creatures'. This was a Buddhist action for acquiring merit;  birds and fish were released immediately after being caught.  To emphasise the character of Wên Hsiian, the account of one of these visits to the tower includes reference to his enjoyment of the spectacle, as all the prisoners jumped to their death.  (A pardon would be given to any prisoner who succeeded in flying from the top of the tower and landing safely).

Apparently these were not the first Chinese attempts at flapping flight. At the beginning of the first century there has been a well authenticated series of experiments involving the imitation of bird motions.  Unfortunately the name of the inventor has not been preserved.  However there is further reference to such attempts in China.  Wang Mang, the only Hsin emperor, found his armies hard pressed during a campaign against the nomadic warriors on the North-West China frontier.  He mobilised all those who professed to have super-natural power, or who were inventors, in order to find new weapons and develop successful tactics to aid him in his war.

"One man said that he could fly a thousand li in a day, and spy out the (movements of the) Huns.  (Wang) Mang tested him without delay.  He took (as it were) the pinions of a great bird for his two wings, his head and whole body were covered over with feathers, and all was interconnected by means of (certain) rings and knots.  He flew a distance of several hundred paces, and then fell to the ground.  Mang saw that the method could be used ..."

Chinese civilisation is also likely to have been the first to develop man-carrying kites.

## THE ADVICE OF POETS

Bladud, the ninth king of ancient Britain, who reigned around 860 B.C., had numerous deeds accredited to him, including the founding of the City of Bath, but his activities were brought to an abrupt end by his interest in flying.  His first and last flight was made,[1] with the aid of a pair of wings, from the roof of the Temple of Apollo in what is now London:

> "Bathe was by Bladud to perfection brought,
> By Necromanticke Artes, to flye he sought,
> As from a Towre he thought to scale the Sky
> He brake his necke, because he soared too high."

The seventeenth century English poet, John Taylor, possibly inspired by the above record, endorsed the sceptics in writing:

> "On high the tempests have much power to wrecke,
> Then best to bide beneath, and safest for the necke."

---

[1] *Geoffrey of Monmouth.  Historia Regum Britanniae, 1147.*

## THE FIRST AUTHENTICATED EUROPEAN ATTEMPTS

An early historically authenticated attempt at man-powered flight was made by the magician Simon in 67 A.D. This attempt is recorded by several religious historians, including Suetonius who claimed that the event was witnessed by the Roman emperor Nero, who was splashed by the blood as the magician crashed to the ground at his feet. A more detailed account of this effort is given in the book *Acta Petri et Pauli*, stating that Simon came to Rome, climbed into a tall tree and jumped, flapping the wings attached to his arms. Initially, he was reportedly successful in maintaining some forward movement, but once he crossed the sacred boundary road close to Nero's villa, his flight was abruptly terminated. According to legend, Simon flew several times, one set of wings he used being more like sails with which, after a preliminary flapping period, he could glide some distance. Schulze and Stiasny[1] consider that to prolong the duration of flight Simon may have flapped his wings, in which case this could be regarded as the earliest recorded man-powered flight to achieve anything close to success. It is unlikely that this conclusion would bear close examination, as many subsequent tower jumpers using the largest practical wing form were unable to approach even the shortest horizontal motion before the inevitable vertical dive.

In the year 875 A.D. it is reported that an Arab, Abul Quasim ben Firnas, known as the Wise Man of Andalusia, attempted a soaring flight by jumping off a cliff, carrying artificial wings tied to his arms. The failure of this flight resulted in his death.

Two centuries later one of the first serious attempts at flight within England took place. The chronicler Henry Knighton, commenting in the fourteenth century on this attempt, stated that in 1060 the Benedictine monk Oliver tried to fly from the tower of the monastery at Malmesbury with the aid of large artificial wings tied to his arms and legs, but crashed and was killed. The 'Flying Monk Inn' at Malmesbury is a continuing reminder of this tragic flight. A second account, in *The History of England*, written by John Milton, leaves one with the impression that Oliver survived his tower jump:

"He in his youth strangely aspiring, had made and fitted wings to his hands and feet; with these on top of a tower, spread out to gather air, he flew more than a furlong, but the wind being too high, came fluttering down, to the maiming of all his limbs; yet so conceited of his art he attributed the cause of his fall to the want of a tail, as birds have, which he forgot to make to his hinder parts."

The Byzantine historian Niketas Akominates tells of an interesting though equally disastrous flight which took place in 1161. It was during the visit of the Turkish Sultan Kilidisch-Arslan II to the Greek King Manuel Komenos in Constantinople that the event took place:

"A Saracen then climbed on to the tower of the Hippodrome and stated that he was going to fly along the race track. He stood on the tower dressed in a very long and broad robe; it was white, and curved willow rods held the fabric taut to form a convex surface. The Saracen intended to fly a short distance in this outfit by trapping the wind in the billows."

---

[1] *H.-G. Schulze and W. Stiasny. "Flug durch Muskelkraft" Naturkunde und Technik, Verlag Fritz Knapp, Frankfurt, Germany, 1936.*

Apparently the King and the Sultan tried in vain to persuade the Saracen to abandon his attempt, which they were sure would end in tragedy, but the Saracen did not heed these warnings. After standing facing the wind for some considerable period, in order to gauge its strength, and ...

"when this appeared sufficient to support him, he leapt to and fro like a bird and seemed to fly in the air. He raised and lowered his arms repeatedly and attempted to trap the wind by flying motions. But his performance did not last long, because he fell to the ground like a stone, every bone in his body was broken, and he died."

Cousin in his *Histoire de Constantinople*, uses the evidence of an eye witness, who tells a story identical in most aspects to that given above, although he suggests that it was the aim of the Saracen to sail through the air 'as a ship'. None of the accounts identify by name the main character in this drama.

One of the most original accounts of a man-powered flight relates to an episode in the life of a Persian architect who had been commissioned by King Shapur I in the tenth century to design and supervise the construction of a tower. The king obviously valued the skill of this architect, for upon completion of the tower the designer was confined to the top of his creation lest his services be hired by others. On the pretext that he required a shelter to protect himself from vultures, the intelligent man obtained some wood and proceeded to construct a pair of wings, with which he escaped to safety.

The Persian was not alone in being imprisoned to preserve the uniqueness of his creation. In his book *Spain*, Sacheverell Sitwell gives an account of the legend of the sculptor of the choir stalls in the cathedral of Plasencia. This craftsman, named Rodrigo Aleman, was imprudent enough to boast that God himself could have done no better work. He was punished for his impiety by being locked in one of the cathedral towers. After a considerable period, during which he managed to keep himself alive by feeding on birds, he fashioned a pair of wings from their feathers and made a fatal jump from the roof.

* * *

*Fig. 1.* Daedalus making his wings – the earliest and
most successful aeronaut?

*Fig. 2.* Icarus and Daedalus are illustrated in what
was the first printed representation of a
flight.

*Fig. 3.* Perseus on Pegasus slaying Medusa. The
ability to fly was widely attributed in
mythology.

*Fig. 4.*   A representation of the legend of King Kawus
who harnessed bird-power.

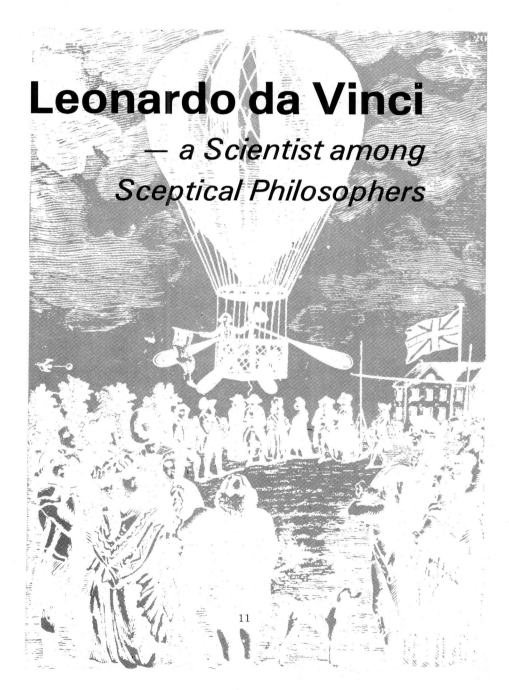

# Leonardo da Vinci
## — a Scientist among Sceptical Philosophers

# 2

The rather lean intermediate period between the mythological and more authenticated accounts of man's attempts to fly is an appropriate time at which to present a few definitions of man-powered flight as detailed later in this book. As will become evident, no definition is currently universally accepted, and most are built up around a particular flight or organised competition, and are thus convenient solely for the purpose of that flight or competition.

## DEFINITIONS OF MAN-POWERED FLIGHT

A simple definition was adopted by the Man-Powered Aircraft Committee at Cranfield in 1957: "Man-powered flight is the act of raising a heavier-than-air craft from the ground, and sustaining controlled flight, solely by the contemporaneous muscular activity of its occupant(s)".

Even this statement was later to be qualified in numerous ways, and the rules of several competitions and the like have become quite complex. The Canadian Aeronautics and Space Institute included minimum altitude and duration criteria, stating that a man-powered flight would take place if the aircraft flew at a height of at least one metre above the take-off point for a duration of one minute.

All of these definitions reject the acceptance of stored energy or assisted launch as part of a man-powered flight. Unaided take-off, ideally the ultimate aim, is included in only one of the three categories of flight proposed by Helmut Haessler, the German man-powered aircraft pioneer, in a letter published in *Flight International* on 19th April, 1962:

"... According to the type of take-off, three separate categories of man-powered flight should be recognised and the performances in each should be recorded separately:

(a)  Flight after unassisted take-off.

(b)  Flight after take-off assisted by externally stored pilot-power. To retain conditions as similar as possible to those of unassisted take-off, the take-off height attained by externally stored pilot-power should be limited to 10 ft (3.05 m). It can be expected that after unassisted take-off, flight distances up to 100 yd (91.4 m) can be achieved with kinetic energy only which corresponds to a theoretical take-off height of about 8 ft (2.4 m).

In both categories the level element of the man-powered flight is inextricably mixed up with the sinking glide element arising from kinetic or potential energy used for take-off. However, the

sinking glide element is kept at the unavoidable minimum and its
influence on the total flight becomes remarkable only in very
short flights.  Therefore, only flights over a distance of more
than three times the distance attributed to the influence of
kinetic or potential energy should be recognised as man-powered
flight in these categories.

(c)  Man-powered flights after take-off with other power (teams
and rubber bungee, car-tow, winch).  Since here the power used
for take-off is outside the limits of categories (a) and (b) only
that part of the flight which is above the height at which
unaccelerated flight condition is attained should be recognised
as man-powered flight.  The part of the flight below this level
will be a glide stretched by man-power.  The flight should be
limited to a maximum height up to which no vertical air movements
could be expected (about 30 ft, 10 m).  At greater heights than
this the flight should be called 'soaring assisted by man-powered
flight'."

Unaided take-off flights have attracted most attention, but several aircraft have flown successfully with assisted launches, and proposals for energy
storage systems have been made by some of the most serious experimenters.  As
the reader realizes the various forms which man-powered aircraft can take, and
the design limitations, he will well appreciate the difficult task facing
those who seek to clarify and extend these definitions to satisfy the
requirements of differing factions throughout the world.

* * *

The four hundred year period from the early fifteenth century to the first
developments in aeronautics as a science were comparatively unproductive as
far as the advancement of man-powered flight was concerned, with one or two
notable exceptions.  This is not to say that attempts to fly did not take
place, and certainly a number of intellectuals found time to offer their
advice and comments to these aeronauts.

During this period the beginning of a new form of development began to
break through, undoubtedly saving the lives of many men.  This was the ability
of designers to draw conclusions as to the viability of their inventions by
calculation and the use of test rigs.  These rigs used mechanical power to
simulate the motion of wings, thus providing information which previously was
obtainable only by jumping off the top of a tower.  Needless to say, only a
very few cautious workers followed what must then have seemed to many a
frustrating waste of time, and the more immediate way was, unfortunately, still
very much to the fore.

## LEONARDO DA VINCI

The most notable exception to the amateur bunglings of the tower jumpers
was Leonardo da Vinci.  He made many contributions to the arts and science,
and in the latter field a considerable proportion of his effort was devoted to
the subject of aeronautics.  Gibbs-Smith, a contemporary aviation historian,
has studied the manuscripts of da Vinci in detail, and considers that much of
his work on aeronautics was influenced by his obsession with man-powered flight
using ornithopters, and his concern with this activity led to his overlooking

some of the more practical aspects of the aircraft he designed[1].  His
prodigous output of scientific literature included about 25 000 words and 500
drawings on the subject of flight.

His first man-powered ornithopter, a single seat machine, was designed
around a pilot operating in the prone position, and is illustrated in *Fig. 5.*
The pilot was secured in the frame by a series of hoops, and used both arms
and legs to motivate the wings.  Early designs of the prone ornithopter
provided for the use of both feet in stirrups operating the wing downstroke,
the wings being returned to their highest position using the hands, operating
two levers.  Other methods which da Vinci suggested for returning the wings
to their upper position included springs and the simultaneous bending of both
knees to pull up the feet.  The transmission system was by means of cords
attached to the feet.  The cords then passed over a pulley to the main
structural member of the wing.  The tip regions of the wings were connected to
the main spar by a further joint, enabling a downward and backward movement
to be imparted to these sections.  Leonardo da Vinci considered that this
action correctly simulated the wing motion of birds.

Leonardo da Vinci was the first person to propose control surfaces for
aircraft, and he installed an interesting control mechanism in one of his
prone man-powered helicopters.  This involved the motion of a cruciform-shaped
rudder-elevator combination, using a head harness.  Transmission of head movements to the rudder was through a shaft attached to a bandage wrapped around
the pilot's head.  Apart from being the first recorded design of a sophisticated control system, the implementation of control using the head is
apparently unique and must be considered as a valid system for a man-powered
aircraft where the pilot requires both arms and legs for propulsive power.

One of da Vinci's last designs for a prone ornithopter incorporated two
sets of wings, the motion being in opposite directions on each side of the
fuselage.  The transmission system utilised cables running over drums, two of
the wing spars being attached to a large rotating disc, adjacent to the top
drum, which were caused to flap, with one wing beating down at the same time as
the wing on the other side of the fuselage was pulled up.  Da Vinci stated
that this machine could also be constructed with a single set of wings, the
arms being used to raise the wings while the feet controlled the downward
motion, as shown in *Fig. 6.*

About the same period that da Vinci designed his prone ornithopters, 1487,
he also did some work on ornithopters in which the pilot was contained in a
boat-shaped fuselage, either in a sitting or standing position.  The wings
were operated by crank handles pivoted about the 'gunwales' on the edge of the
fuselage.  A moveable tailplane of considerable size was also included.  An
important feature lacking in this design was a wing motion to provide forward
thrust, as opposed to the vertical flapping motion with which it was hoped to
generate lift.

The short bibliographical work by Gibbs-Smith on da Vinci's aeronautical
achievements contains information on his standing ornithopter designs, on
which he concentrated towards the end of the 1480s.  It is stated that da
Vinci's emotional influences on his aeronautical work reached their climax
during this period, and one must agree with this conclusion in studying the

---

[1] *C.H. Gibbs-Smith. 'Leonardo da Vinci's Aeronautics'. HMSO, London 1967.*

practicability of his standing ornithopters.  The pilot stood in a bowl-shaped craft and operated four beating wings using a rather complicated transmission system which utilised both arms and legs and in practice would have proved to be exceptionally heavy.  Even the head is used, with neck and chest muscles, to provide effort, and da Vinci writes:

> "This man exerts with his head a force that is equal to two hundred pounds and with his hands a force of two hundred pounds, and this is what the man weighs.  The movement of the wings will be cross-wise, after the manner of the gait of a horse.  So for this reason I maintain that this method is better than any other."

Less advanced systems were the subject of patent applications as late as the 1880s[1].

A foremost feature of this ornithopter was a retractable undercarriage, consisting of a ladder and third leg.  At the top of the ladder a trap door led into the bowl fuselage.  On take-off it was proposed to raise the ladder and stabilising leg until they were flush with the bowl base.  Describing the size of the machine, da Vinci refers to the undercarriage:

> "Ladder for descending and ascending;  let it be twelve braccia high, and let the span of the wings be forty braccia, and their elevation eight braccia, and the body from stem to prow twenty braccia, and its height five braccia, and let the outside cover be all of cane and cloth ... Make the ladders curved to correspond with the body."

More realistic designs followed, and in the period 1497 - 1500 da Vinci worked on a machine in which only the wing tips flapped, the main body of the wings being fixed.  The pilot was slung between the wings but did not apparently use the principle of centre of gravity shift for control of the aircraft as did Lilienthal in the 1890s.  This design, shown in *Fig. 7*, was one of his most advanced ornithopters, and appears to be his final attempt at design exercises involving complete aircraft.  He later concentrated on attempting to perfect wing and control surface design, and in 1505 wrote a paper on bird flight, this having a considerable influence on his wing design.

After considerable study of the documents existing on Leonardo da Vinci's aeronautical work, Gibbs-Smith concludes that it was very unlikely that da Vinci built one of his ornithopters, and in studying the design of these machines, one must conclude that their weight alone would be sufficient to completely prevent flight.  Gibbs-Smith concedes that a full scale mock-up of one of the machines may have been completed, but he has considerable reservations, influenced not least by the fact that Leonardo da Vinci tended to leave many projects incompleted, moving into new and more interesting fields of study.

## PHILOSOPHICAL ALLUSIONS

The eminent thirteenth century philosopher Friar Roger Bacon also turned his hand to the problem of flight.  It cannot be ascertained from his writings that he ever realised any of his grand projects, and he concludes his treatise c.1256 thus:  "There may be some flying instrument, so that a man sitting in

---

[1] *See Appendix I.*

the middle of the instrument, and turning some mechanism, may put in motion some artificial wings which may beat the air like a bird flying".

Soon after Bacon's time, Hatton Turner in his *Astra Castra* states that projects were instituted to train infants in the exercise of flying with artificial wings, based on the ideas of the philosophers and artists of that day.  If we credit the accounts of some of their experiments, it would seem that considerable progress was made in that way.  "The individuals who used the wings could skim over the surface of the earth with a great deal of ease and celerity.  This was accomplished by running and flapping the wings with the arms."

Wise, in his *History and Practice of Aeronautics*[1] suggests that this form of flying, in conjunction with a small balloon (hydrogen) which could contribute partially to the required lift, would be extremely useful, and one could envisage 'jumps' of the order of one hundred metres, aided by balloon and flapping wings.

Cuperus, in his treatise on the *Excellence of Man*, contends that the faculty for flying by the use of artificial wings can be attained.  However Borelli, a Neapolitan mathematician, refutes the idea.  Borelli's studies on flight were conducted towards the end of the seventeenth century, and in his thesis *De Motu Animalium*, he determined the extent to which the motive power of the pectoral, or chest muscles would have to exceed the drag of the body in order to enable man to achieve flight.  His comparisons of man and birds led him to conclude that man-powered flight by means of flapping wings would not be feasible unless the muscular bulk of man were to increase, or his weight to be substantially reduced with no corresponding loss of power.

Wise discovered a letter, which was published soon after in some of the French scientific journals, together with the copy of an address presented to the King of Portugal in 1709 by the Friar called Bartholomew Laurence de Gusman. In this the petitioner represents himself as having invented a flying machine capable of carrying passengers and navigating through the air very swiftly. He also, in effect, requested a patent, upon which the King issued the following order:

"Agreeable to the advice of my council, I order the pain of death against the transgressor,  And, in order to encourage the suppliant to apply himself with zeal towards improving the machine which is capable of producing the effects mentioned by him, I also grant unto him the first vacancy in my College of Barcelona, with the annual pension of 600 000 reis during his life."

"The 17th day of April, 1709."

The machine is reputed to have resembled a bird.  The wind was used to inflate a form of sail, otherwise bellows were to be used.  Apparently the lack of expertise shown by de Gusman made his receipt of a professorship most unlikely.

Bourgois, in his *Récherches sur l'Art de Voler* considers that de Gusman

---

[1] *John Wise 'History and Practice of Aeronautics'  Joseph A. Speel, Philadelphia, 1850.*

and Bartholomew Laurence were distinct people, the latter being granted the 'patent' above.  De Gusman was reputed to have worked on some form of air balloon in 1736.

Towards the end of the fifteenth century, about the time of Leonardo da Vinci's work on ornithopters, a fellow countryman and scholar, John Baptist Danti of Perugia, attempted a flight from a prominence near Lake Trasimene. Details are recorded in the historical records of the city of Perugia:

"One day when many distinguished people had come to Perugia for the wedding ceremony of Paolo Begliono;  and were holding a festival in the main street;  Danti suddenly jumped from a nearby tower using a rowing device with wings, which he had constructed in proportion to the weight of his body.  He flew successfully over the market place, producing a fair amount of noise, and was watched by a large crowd.  But when he had flown a distance of barely three hundred paces, an iron component on the left hand wing broke, so that he fell onto the roof of the Church of Maria delle Virgine, and was seriously injured."

Also, in 1496 we read that an old cantor of Nuremburg, named Senecio, broke an arm during an attempted flight.

*  *  *

The excuses put forward by unsuccessful aviators were in many cases so ridiculous as to be extremely humorous.  John Damian, an Italian by birth, came to Edinburgh in 1501 for installation as Abbot of Tungland by decree of James IV.  Declaring that he would be able to overtake an ambassador then on his way by sea to France, he launched himself from the wall of Stirling Castle, using a pair of feathered wings.  After the inevitable vertical dive, which fortunately resulted only in a broken thigh-bone, he commented that the failure was no doubt due to the fact that hen feathers had been included in the fabrication of the wings, and so these had been attracted towards a local barnyard, rather than up into the sky!

## JOAO TORTO AND KASPER MOHR

An early claim for successful flight in an heavier-than-air machine is chronicled by the Portuguese writer Donna Maria da Gloria.  He reported that Joao Torto, a school master by profession, flew in the year 1540 at Vizeu, the capital of the Portuguese province of Biera Alta.

The machine had four calico wings, the lower pair being slightly smaller than the upper set.  All four wings were connected together by means of fabric-covered iron rings, and the pilot thrust his arms and legs through these and carried the flying machine on his back.  A concession to the acknowledged superiority of birds in achieving flight was made by Torto in his flying suit design, the main feature of which was a hood on which was painted the open beak of an eagle.  The aperture in this beak was intended to provide the pilot with a view of the ground.

It is reported that on June 1st, 1540, Torto sent the city herald through the streets of Vizeu with the message "I proclaim to all the inhabitants of this city that before this month comes to an end, a very great miracle will be seen, in the form of a man who will fly with magic wings from the clock tower of the cathedral towards Matthew's Fields."

    As predicted, preparations for the flight took place on June 20th of that year.  The machine was hoisted onto the roof of the belfry.  Beneath the tower the citizens of the town and many peasants from surrounding regions of the province had collected in such great numbers that "no drops of water could have reached the ground".  Torto threw himself into the air as the clock in the tower struck five.  "As if a miracle really had come to pass", wrote the chronicler, "Master Torto made his way through the air in the direction of Matthew's Fields."  The flight, however, was short lived.  The hood through which the flier was able to see his path, slipped, and he momentarily was unable to see his position and, in his confused struggles, crashed on to a roof. Although he was not seriously injured, there is no report of a further attempt at flight.

    Donna Gloria omits the significant details appertaining to the duration of flight and distance flown, and there is no evidence to support claims that Torto was more successful than his contemporaries and followers in achieving a prolonged flight.

    A second person of the period who aspired to man-powered flight was the Swabian monk Kasper Mohr, of Schussenried, in Württemberg.  He lived from 1575 to 1625 and as well as being a renowned theologian, was reported to be competent in the fields of mathematics, sculpture, painting and mechanics. His aim was to simulate bird flight by making wings of goose feathers, attached to his body by means of whip thongs.  Fortunately he was prevented by his superiors at the last moment from jumping off the roof of the monastery at Prämonstraten.  A painting of the monk preparing for his flight can still be seen in the library of the monastery at Schussenried.

## BISHOP JOHN WILKINS

    John Wilkins, Master of Trinity College, Cambridge and later Bishop of Chester, in his 'Opus Magnum', *Mathematical Magick*, which was published in 1648, has a section devoted to his thoughts on various forms of flight, which he believed would best take place at high altitude, outside the gravitational effects of earth.  The problems of reaching such a height would be solved in one of four ways:

            "By Spirits or Angels.
            By the help of Fowls.
            By wings fastened immediately to the body.
            By a flying Chariot."

Wilkins was doubtful as to the merits of the first three techniques, citing a number of examples of tower jumpers who sustained injury.  He favoured the flying chariot, in which one or several crew members would work to aid propulsion.

    A friend and contemporary of Bishop Wilkins, Robert Hooke, is actually reported to have made small flying models in 1655, and to have discussed the power and sizes of muscles with Francis Willughby, the ornithologist. Willughby likened the strength of a bird's pectoral muscles to a man's legs rather than his arms, concluding that a man would have to use his leg muscles if he hoped to fly.  These comments were made generally known in his

*Ornithologiae Libritres*, published in 1676 after Willughby's death.

Bishop Wilkins could well have influenced Willughby in his investigations, for the former is known to have stressed the importance of larger muscles, using the domestic fowl as an example where the wings (arms) are too weak for fast continuous flapping. Needless to say, little notice was taken of this advice, although it was to receive supporting evidence from many other sources.

Hooke was no doubt gratified when he heard that Bernier, who lived in Frankfurt, Germany, had constructed a set of wings actuated by both arms and legs. In his *Philosophical Collections*, compiled in 1679, Hooke describes how Bernier tested his design by jumping in turn off a stool, table, ground floor window, second floor window and garret. Although Hooke's source states that the flights were successful, other accounts conclude that Bernier suffered multiple fractures.

* * *

One of the most well known of the French 'aeronauts', Besnier, a locksmith, designed and constructed a pair of paddle-like devices, which he pivoted about each shoulder. At the end of each paddle arm were V-shaped planes, actuated by hands and feet, the paddles being connected to the latter by cords. Following reportedly successful flights, Besnier sold his blades to a showman who was killed during his first professional exhibition after jumping from a height of about 15 metres, *(see Fig. 8)*.

Also in France, a tight-rope walker named Allard attempted a flight before King Louis XIV at St. Germain. Allard affirmed that he would fly from the terrace towards the woods of Vesinet. With wings strapped to his arms, he launched himself in the promised direction, crashed to the ground and was seriously injured. The attempt was reportedly intended to be more in the nature of a glide than a flapping flight.

Three other comparatively insignificant attempts were made during the seventeenth century. A painter named Guidotti attempted a flight at Lucca c.1628, and in Nuremburg, a mechanic, Hautsch, built a flying machine with which he tried to take off (unsuccessfully) from level ground, rather than jumping from a tower. Less well-advised was Bolori, an Italian clockmaker who launched himself from the roof of Troyes Cathedral in France.

Although the 1600's were remarkable in that the attention of some of the reputable scientists and philosophers of the period was drawn towards considering the problems of flight, the machines constructed during this century were in general developed by people with no appreciation whatsoever of the problems involved. Communication between the stratas of society were of course, so poor that the works of, for example, Willughby, were unavailable to the back-yard inventor, and more often than not, the latter would have found them unintelligible in any case.

## 'INCENTIVES TO FLY'

One of the longest man-powered flights, of several kilometers, was claimed on behalf of Hezarfen Celebi, who in the seventeenth century flew from a tower in Galatia on the Bosphorus, to Scutari.

Some of the encouragement that these would-be fliers received was such that one could say for certain that their failure was not for lack of enthusiasm. There is, for example, an account of a Russian peasant who tried to fly in 1680, but was whipped for failing to take off. Other misfortunes were more numerous. Somewhat later, Canon Ogar of Rosoy Abbey fell into a bush during an attempt to fly, and an unlucky (or lucky!) priest from Peronne dived into a moat following his abortive efforts.[1]

A passage in a book telling the fortunes of a farmer named Jocosa, written in 1685, reveals a rather sceptical attitude towards man-powered flight, although contemporary thought no doubt viewed the piece as a most realistic assessment. The paragraph reads as follows:

"How often the fellows must have laughed at those who wanted to make men fly. Because they did not know that it is written 'As the bird is born to fly, so is man born to work (sic);' and what would man finally have achieved if he could fly? If he actually succeeded in doing such a thing, he would find therein not only few uses but great inconveniences, which would be caused by dangling back and forth. Such an inventor would certainly be the cause of many casualties. For supposing that one could bring it to pass, all journals would truely say how this man and that had fallen from the air to his death and they would have to lift the man either out of rivers or off spiked fences to bury him half chewed up."

The seventeenth century is perhaps best summed up with a light-hearted reference to man-powered flight made by the Marquis of Worcester in his *Century of Inventions*, written in 1663: "How to make a man fly; which I have tried with a little boy of ten years old in a Barn, from one end to the other, on a haymow."

* * *

The eighteenth century saw no let up in the efforts of the tower jumpers to defeat gravity, but a certain change in the nature of attempts was detectable, although at first no success was recorded.

In 1742 the Marquis de Bacqueville, portrayed in *Fig. 9*, attempted to fly the river Seine in Paris with the use of wings strapped to his arms and legs. He jumped off the roof of a house overlooking the bank and crashed onto a passing barge, breaking his legs.

Aviators were not encouraged by the words of William Cowper, who in 1783 wrote: "If man had been intended to fly, God would have provided him with wings." This was, of course, not the first time such a sentiment had received publicity, and it was certainly not the last.

## FRIARY CYPRIAN

The history of flight in Hungary by all accounts was initiated by Friar Cyprian (1708 - 1775). Friar Cyprian was born in northern Hungary, and much of his early life was spent at the Vörös Klastrom (Red Monastery) near

---

[1] *H-G Shulze and W. Stiasny 'Flug durch Muskelkraft'. Frankfurt A.M. Naturkunde und Technik, Verlag Fritz Knapp 1936 (Translated from the German)*

Sublechnic in the Szepes Province.  During his period of study at the Monastery,
he was under the direction of the Carmaldolese Order, which he was later to
join subsequent to working as an assistant to a physician in Löcse, and
travelling.  He joined the Carmaldolese Order in 1753, being employed as the
friar's physician, barber, chemist and cook.  Friar Cyprian had a strong
interest in botany, and his collection of herbs included 272 specimens from
the Tatra region, with references to the medical use of each.  The collection
is still preserved in the Museum of Poprad.  He seems, like his predecessor
Kasper Mohr, to have had numerous interests, and an inquisitive mind trained in
scientific disciplines, and his interest in flying is not inconsistent with his
general farsightedness.

Engineer Reissig, who lived in Szepes Province around the beginning of the
nineteenth century, wrote:  "The friar had been working on two wings in the
monastery's laboratory for years.  These he fastened to his arms.  They had
extensions which could be moved through a clock-like apparatus by the main
wings".  In all probability the wings were based on the bone structure and
muscle system of the bat, slightly bent and covered by thin animal skin.  The
extensions at the wingtips could be controlled by straps fixed to the legs.
Before he attempted to fly in the open air, he carried out some trials.
According to available sources, these took place in 1768.  In Samuel Bredetzky's
*Topographisches Taschenbuch für Ungarn* (Hungary's Topographic Handbook),
published in 1802, the event was also mentioned:  "Solitude may affect the
spirit of invention favourably.  Before the world heard of aerostats, experiments were made here by a friar called Cyprian, who lifted himself into the
air with the help of wings".  The thirty-fourth issue of the periodical
*Hasznos Mulatságok* in 1825 includes the following passage:  "Cyprian was the
first to experiment to fly with artificial wings in the Carpathians.  Actually
he did more than that:  in the open air he flew high up from the Red Monastery
to the far away Mount Korona".

His superiors considered Cyprian's flying activities as blasphemous, and
his flying machine was burned by order of the Bishop of Nyitra.  One source
claims that he escaped punishment only because Joseph II dissolved all religious
orders in his Empire;  if, however, Cyprian really died in 1775 this is
incorrect, for Joseph's decree was published in 1781.  In any case, his life is
little known beyond his flying experiments.

Aviation writer and air force pilot F. Hefgy visited the locality in 1933,
and came to the conclusion that some kind of flight, or rather descent, would
have been possible.  He favours the theory that the large surface of wings was
used as a parachute, while others believe that Cyprian glided down from Mount
Korona.

The fact that the device was fixed to Cyprian's body and that he planned
to manipulate, if only to direct, his wings while in flight would probably
classify his attempt as a man-powered flight, even if he used himself as a
glider, or was compelled to glide.

* * *

While the extremely simple gliders or flapping wing machines, generally
consisting of some crude form of lifting surface strapped to the arms, were
made in considerable numbers (of course many of these would never be recorded
and will always remain unknown), the trend towards a more sophisticated
approach which began, or perhaps more correctly, was reborn in the seventeenth

century was gaining momentum.  This was partially evident in the background of
the people who were beginning to involve themselves in the problems of flight
in general.  It also became manifest in the designs produced by the inventors.
No longer did one seem to be satisfied with a simple pair of bird-like wings -
landing gear, control surfaces, and even fuselages were shortly to become the
*sina qua non* of any self-respecting designer.

   Not all designs of course, reached the construction stage, nor in some
cases was this the intention of their creators.  In 1768 for example, the mathematician A.J. Paucton, in his work *Théorie de la Vis Archimèdes* suggested a
man-powered helicopter, with two helical screws, one for lift and one to aid
forward propulsion.  Although never constructed, this machine appears to be the
first reference to a man-powered helicopter.

   Earlier, in 1722, Abbé Desforges of Etampes designed and built a 'flying
carriage'.  This was a wickerwork basket fitted with wings.  The main feature
of the machine was a large canopy of fabric above the wings, used as a form of
para-wing.  The movable wings, however, proved unable to generate sufficient
lift;  on the first trial four men held the basket aloft while the Abbé flapped
the wings, to no avail.

   Another unusually complex construction was revealed in a collection of
diagrams discovered in 1921 in the Staatsarchiv of Thuringia at Greiz.  These
show a man-powered aircraft proposed by Melchior Bauer during the early 1760's.
The aircraft was to consist of a bogey-like platform running on four wheels,
two of which were driven by the pilot, standing on the bogey.  The wing
arrangement, in effect a biplane, comprised a fixed upper surface, supported
by a central strut to which were attached a number of bracing wires.  Construction was canvas based on a wood lattice framework, and the span of the upper
wing was approximately 6 metres.  Propulsion was obtained from a rocking system
of auxiliary beating surfaces mounted below this wing, so connected as to form,
when stationary, another almost continuous surface.  This flat valve operated
in such a way that when the port side of the frame was rising, with its flaps
open and thus having a low drag and producing little negative lift, the
starbord wing was moving downwards with its flaps shut, creating a thrust with
a component of force in the rearward direction.  Reports suggested that this
aircraft was constructed, but never flew.


## MEERWEIN

   Karl Friedrich Meerwein, architect to the Prince of Baden and a resident of
Karlsruhe, in his pamphlet *L'Art de voler à la Manière des Oiseaux*, published
in 1784, describes how he calculated, aided by some data obtained by experiments
on ducks, the wing area necessary to support a man.  The value of 11.7 square
metres obtained by Meerwein was verified later by Otto Lilienthal for the case
of a machine with an all-up weight of 91 kg.  Not satisfied with a solely
academic approach to the subject, Meerwein proceeded to construct an aircraft,
having a span approaching 10 metres (illustrated in *Fig. 10*);  this was
strapped to his back and the pilot used his arms to push a bar pivoted at the
centre of each wing.  Motion of the bar caused flapping of the wings.

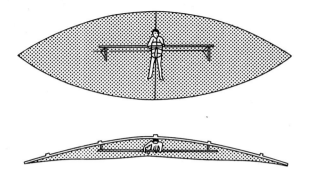

*Fig. 10* Karl Friedrich Meerwein's Ornithopter, dated 1782.

## BLANCHARD

A further step was taken by Jean-Pierre (Francois) Blanchard in the construction of an amphibious ornithopter, called by Blanchard the *Vaisseau - Volant*. His philosophy of flight: "It is not the material or form of wings which causes flight, but the volume and celerity of movement, which should be as rapid as possible" could not be considered as an encouraging start, and only served to warn others of failure. However, his work is of some interest, as he was then almost alone in appreciating the advisability of using both arms and legs to obtain the maximum amount of power.

Blanchard's machine had four wings, mounted in a light-weight car resembling a boat with a canopy. The hand and foot levers were used to operate the wings, but take-off was to elude the designer. Later in 1782 he decided to construct a man-powered helicopter, again without success.

It is worth noting here that Blanchard was one of the pioneers of the balloon, invented in 1793. His most significant contribution to aeronautics was his use of a propeller, manually rotated, mounted on a balloon. Using this, he was able to exert some limited control on the speed and direction of flight. A more ambitious design for an airship propelled by three manually operated propellers was proposed by General Meusnier in the following year. This machine was never constructed, but a contemporary balloon is illustrated in *Fig. 11*.

Renaux and Gérard are both credited with ornithopter designs in 1784, and an Italian, Titus Buratini, is reported to have constructed two prototype machines. Gérard's work is interesting not for its use of man-power, which he regarded as being of secondary importance, but for the development of a motor driven by detonations, and the incorporation in his machine of a rudder and cushioning undercarriage.

## LITERARY RERERENCES

The eighteenth century was remarkable for a profusion of works of fiction containing passages on fantasies of flight. More serious articles began to appear, in which eminent members of society forecast the types and uses of aircraft, once they were capable of flight.

Among those making such references to aviation are Joseph Addison in the *Guardian* (1713), Swift in *Gulliver's Travels* (1726), Samual Brunt in *A Voyage to Cacklogallinia* (1727), Robert Paltock in *The Life and Adventures of Peter Wilkins* (1751), Dr. Johnson in *The Rambler* (1752) and Restif de la Bretonne in *La Découverte Australe* (1781).

Robert Paltock, who could be described as an early science fiction writer, saw Peter Wilkins as the young hero among an aerial race whose wings were an appendage to their bodies. These wings, shown in *Fig. 12*, were used for swimming as well as flying, and were 'retractable'.

In addition to his literary pursuits, Dr. Johnson seems to have taken a more direct interest in flight. Writing to Mrs. Thrale in 1784, he refers to the construction of a flying machine then in progress.

"A daring projector who, disdaining the help of fumes and vapours, is making better than Daedalean wings with which he will master the balloon and its companions as an eagle masters a goose..... a subscription of £800 has been raised for the wire and workmanship of iron wings - one pair of which, and I think a tail, are now shown at the Haymarket, and they are making another pair at Birmingham. The weight of the whole was to be about 200 lb."

No other records to assist in the identification of this craft have been found.

* * *

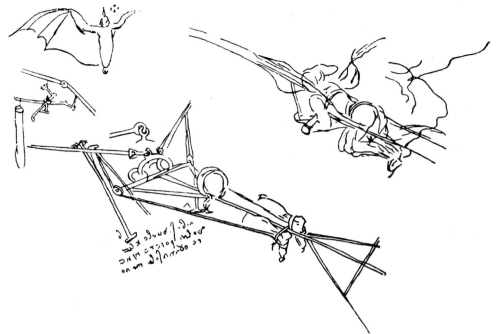

*Fig. 5.* Drawings illustrating the action of da
Vinci's earliest ornithopters. c. 1487.

*Fig. 6.* A prone ornithopter of Leonardo da Vinci.

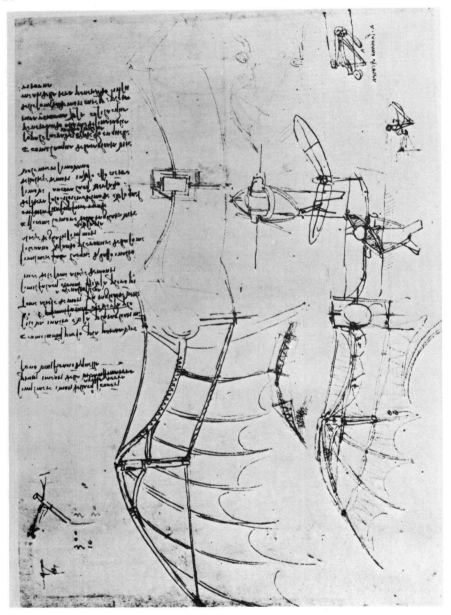

*Fig. 7.*  A drawing by Leonardo da Vinci, showing a fixed wing design with ornithopter-type wing tips.

*Fig. 8.* The novel lifting device proposed by
Besnier in 1673.

*Fig. 9.* The Marquis de Bacqueville attempting to
cross the Seine. He unfortunately fell
heavily into one of the small boats.

*Fig. 11.* Ascent by Lunardi in his balloon, 15th
          September 1784. The 'oars' could be used
          for guidance and a crude form of propul-
          sion.

*Fig. 12.* A scene from 'The Life and Adventures of Peter
Wilkins', as portrayed by Robert Paltock in
1751.

# The Nineteenth Century

## – *The Zenith of the Ornithopter*

*3*

The nineteenth century, as far as flight in general, and man-powered flight in particular, was concerned, commenced badly – in 1801 a French General, Resnier de Goné tested an ornithopter which he had begun designing thirteen years previously, but succeeded only in making a dive into the river Charente. Undeterred, the General made a further attempt over land, but unfortunately broke a leg. He was seventy-two years of age at the time.

## SIR GEORGE CAYLEY

Sir George Cayley, acknowledged by many to be the inventor of the aeroplane, established the basis for modern aerodynamics and made many contributions to aeronautical science. He was born in Scarborough in 1773 and his experiments with simple models began twenty-three years later. One of his first designs for a full size aircraft, conceived in 1799, showed a marked breakaway from contemporary thought in that the main wings were fixed.

This separation of the devices producing thrust and lift, normally unsuccessfully combined in the ornithopter in the form of a flapping wing, was very significant. A silver disc engraved by Cayley and illustrated in *Fig. 13*, shows the forces produced by a wing on one side, and a sketch of his proposed aircraft on the other. The aircraft is a monoplane, with the pilot sitting in a boat-type fuselage. The tail unit combined both rudder and elevator in a cruciform, and the operation was via a tiller in the cockpit section. It was propelled forward by oars pulled by the pilot in rowing fashion. A number of pilot positions were provided so that the pilot did not have to concentrate on keeping the oars in a horizontal position.

Cayley cannot be described as a champion of man-powered flight – in fact most of his efforts at designing man-powered aircraft were the result of frustrating attempts to find suitable mechanical power plants for his machines, having considered explosives, steam engines and a hot air engine which he invented. Such was his objectivity, with one noticeable exception, in his approach to flight he soon realised that without a power plant of some considerable output, any sustained flight would be impossible with the technology available to him at that time. The exception was his reported faith[1] in the story that Degen had flown in Vienna by man-power alone[2]. Cayley used his evidence of Degen's flight to support his arguments and faith in the future of aviation. Most misleadingly, no mention was made of the balloon used to support most, if not all, of Degen's weight during his wing-flapping exercises,

---

[1] *C.H. Gibbs-Smith 'Sir George Cayley 1773-1857' HMSO, London 1968.*

[2] *Degen's flights took place during the period 1808 – 1812, after much of Cayley's formulative work on fixed wing aircraft, and the influence of these flights on Cayley's work does not seem too great.*

although this is also omitted from contemporary drawings *(Fig. 14)*.

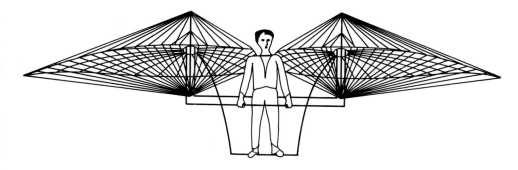

*Fig. 14.* Jacob Degen's ornithopter, conveniently represented in old prints minus the supporting balloon.

Helicopter design was also a feature of Cayley's work. Following the construction of his model helicopter in 1798, utilising two contra-rotating rotors consisting of four feathers, he speculated on full scale helicopters: "If, in lieu of these small feathers, large planes, containing together 200 square feet, were similarly placed, or in any other more convenient position, and were turned by a man, ... a similar effect (flight) would be the consequence, and for the mere purpose of ascent, this is perhaps the best apparatus."

In a paper on aerial navigation, published in Nicholson's Journal in 1809, Cayley mentions the possibility of utilising man as a source of power, as an alternative to other power units. At this stage he was still sure that the wing area should be 200 square feet (18.6 m$^2$).

Although Cayley's first aircraft designs had been monoplanes, he later expressed a strong preference for multiplanes, mainly on the basis that a monoplane of sufficient span would be structurally inferior to a biplane or triplane arrangement. An example of this later work was a triplane constructed in 1849. It had a wing area of 31.4 m$^2$, and an empty weight of 59 kg. The fuselage was similar to his earlier man-powered aircraft design, with a tiller-operated cruciform tail unit. Above the fuselage, supported on four struts, was the triplane wing assembly, notable for the close spacing of the wings. Propulsion was by short flapping wings situated between the fuselage and main lifting surfaces. A short flight was recorded using this machine - by towing against a breeze a young boy was floated off the ground for a few metres, although it is doubtful if the flapping mechanism made any contribution to the flight.

We will see later that the pilot accounts for a very large proportion of the weight of current man-powered aircraft, mainly due to advances in fabrication techniques using light materials. It is, therefore, surprising that no serious construction group working on man-powered flight has, apparently, considered the use of an intelligent child of, say fourteen years of age, or a dwarf, as pilot. Further reference is made to this later.

Man-powered flight was almost an incidental part of Cayley's work, but he deserves mention here, if only for his invention of the fixed wing aircraft, which much later, was to be the basis for the more successful man-powered aircraft.

## JAKOB DEGEN

Among the first men in Austria to concern themselves with the construction and testing of a man-powered aircraft was the Vienna clock-maker Jakob Degen in whom Sir George Cayley had such faith. He was born at Oberwil in the canton of Baselland. The record of births and baptisms of the parish contains an entry that Jacob Degen was born there on 14th November, 1761, although other references give his date of birth in 1756. Initially he was destined to learn the trade of ribbon-maker, which was a popular local industry. Degen's preoccupation with mechanical systems induced him in 1775 to take up the trade of clock-maker, while maintaining a broad interest in other aspects of engineering.

His attempts at man-powered flight were made between 1808 and 1812. He designed and constructed an ornithopter with a wing area of 12 m$^2$ resembling two large canopies supported by a number of wires. Degan demonstrated his machine in Vienna, Berlin, and later, in Paris in 1812. Initial tests were carried out by measuring the amount of lift he could generate against a counter-weight, but later his flying machine with taffeta-covered wings (the material being loosely attached to simulate the 'free action of bird feathers') was suspended beneath a small hydrogen balloon. During an early flight in Vienna, Degen is reputed to have risen 16.5 metres[1], but in October 1812, during an attempted ascent in Paris, Degen was attacked and injured by the spectators, no doubt because the use of the balloon was considered an unfair aid to man-powered flight. (See also *Fig. 22*).

* * *

At the same time as Degen was experimenting with balloons, Albrecht Berblinger was constructing a less successful ornithopter, with which he attempted to fly in 1811. Berblinger's machine, the wings of which the pilot must have had great difficulty in supporting, much less flapping, had a span of about 5 metres. The wings were of circular planform with a vertical strut containing bracing wires passing through each centre. A tail was provided but no control was envisaged. On his first flight, on 30th May, 1811, he crashed into the river Donau.

Although most of the attempts of this period took place on the continent of Europe, interest in man-powered flight was still active in Britain. The English painter Thomas Walker published, in 1810, a book entitled *A Treatise on the Art of Flying by Mechanical Means*, in which he explained the principles, as he saw them, of bird flight. He used this knowledge to present drawings and instructions for the manufacture of a man-powered ornithopter, the pilot using his arms and legs to move the wings. Walker, unlike his contemporary, Sir George Cayley, did not construct a prototype.

---

[1] *A vertical flight is of no significance in this case, and no authoritative reference to Degen's use of wings to obtain forward motion in any direction other than that of the wind at the time has been found.*

The thirty year period up to the end of Cayley's work saw many fruitless attempts to defeat gravity, one of the most ambitious of which was the system proposed by the French Count, Adolphe de Lambertye.  He designed an ornithopter and suggested using a helicopter to ferry passengers between the vessel and the ground.  One can sympathise with the crew of the ornithopter!

In 1825 Vittorio Sarti, an Italian, designed a helicopter with a double airscrew in which vanes of the windmill were hinged, and could flap to set up a forward motion.  The machine proved too heavy for man-power, and the lack of an alternative light engine caused Sarti to abandon the project.  Three years later a helicopter designed and constructed in England by David Mayer, a carpenter, also failed to take off.

Ornithopters remained popular.  Duchesnay constructed one in 1845, and in the same year a design was produced by von Drieberg.  A year later Marc Seguin joined a growing list of failures with his ornithopter.  Earlier, in 1843 Dr. Miller of London designed a man-powered ornithopter in a more serious attempt to simulate bird flight, the wings being actuated by levers and cables worked by both hands and feet.

## A SELECTION OF ORNITHOPTERS

In 1854 Bréant again demonstrated his faith in the ornithopter by suggesting that manual depression of the wings by the pilot would best be aided by raising them with elastic, (see *Fig. 16*).

A sea captain from Brittany, Jean-Marie le Bris, constructed an 'artificial albatross' in 1855.  Illustrated in *Fig. 17*, the machine had a body of light wood in the form of a skiff, and supported hinged wings.  With an overall span of 7 metres and a wing area of over 18 $m^2$, the effort needed to flap the wings must have been considerable.  The wing structural framework was made of light ash, covered with flannel and tensioned steel wires.  A system of cords and pulleys was used as an aid to flapping, and in 1857 a single attempted flight was made with the assistance of a horse towing the machine on a trolley.  In 1868 a second machine, similar to Le Bris' first prototype, was tested. Apparently several glides were made, ballast being carried in lieu of a pilot. This machine was eventually destroyed in a crash before any successful man-powered flights were made.

An Alsation businessman, J.J. Bourcart, constructed an ornithopter with two pairs of wings pivoted on his shoulders, and seemingly propelled by a treading motion of the feet using wire slings.  Although Bourcart claimed encouraging results, his aircraft, shown in *Fig. 18*, is assumed to have been unsuccessful.  However, Bourcart was one of the first people to financially encourage aviation, offering prizes totalling 7,000 francs for flights of five and twenty minutes duration.  His prototype ornithopters numbered three in all, and were built between 1886 and 1910.  The family association with heavier-than-air flight had extended over one hundred years.

In 1864[1] Struvé and Telescheff suggested a man-powered ornithopter with five pairs of wings, in the hope that this would overcome the problem of stability.  Also in 1864, Claudel patented a machine which consisted of a

---

[1] *In a patent dated 1864 the design is attributed to a certain Marc Mennons (See Appendix I)*

combination of fixed wings and longitudinally rotating surfaces which were to twist into propellers.

The German, Otto Lilienthal, (see Chapter 4), was becoming interested in ornithopters around this period, and in 1867 he constructed a six wing orni-thopter and tested this against a counterweight, the total weight of machine and pilot being 80 kg. However, it was left to an Englishman to return to a 'fixed-wing' machine, pioneered by Cayley. In 1868 Charles Spencer made a set of wings which were to be attached to a man. Instead of flapping these wings, which had a total area of 10.2 m$^2$, small auxiliary propellers were fixed at the tips and driven by man-power. It is reported that this machine had limited success, more likely as a glider, and Spencer subsequently started on the design of a delta winged aircraft, similarly propelled. One of his machines was exhibited in a partially completed state at the exhibition arranged by the Aeronautical Society[1], founded two years previously, and held at the Crystal Palace in 1868.

## THE AERONAUTICAL SOCIETY — A LEARNED FORUM

The Aeronautical Society, with the Duke of Argyll as President, provided at the time a unique platform for discussion on topics concerned with aviation, and, not surprisingly, many papers dealing in one way or another with man-powered flight were read at its meetings.

In the Annual Report of the Society for its first year of operation, 1866, notes are given concerning a paper read by M. Henri Réda of St. Martin, entitled *A Projected Experiment in Aerial Locomotion*. The author proposed to attach to his shoulders a pair of wings constructed with ribs covered with silk or some other light fabric. For obtaining the required speed to enable the pilot to take off "by the inclined surface of the wing against the air, prolongations were to be connected with the feet in the form of stilts". It was the aim of the inventor to use the stilts to allow him to travel over the land at high speed, each step being associated with a jump, resulting in a short flight. He could also "cross rivers and lakes with the same ease".[2]

Following the presentation of this paper, a discussion took place during which a questioner asked what form of motive power could be used. The author explained that he intended his invention to be used in conjunction with a running motion on the part of the wearer. The Chairman of the meeting rather doubted M. Réda's claims.

During 1867, F.W. Breary, secretary of the Aeronautical Society, proposed that an *Exhibition of Machinery and Articles connected with Aeronautics* should be held. The exhibition was duly opened at Crystal Palace on the 25th July, 1868, and lasted eleven days. A general view of the exhibition floor is shown in *Fig. 19*.

---

[1]  *The Aeronautical Society is now the Royal Aeronautical Society, and is res-ponsible for administration of the current Kremer Competition, in conjunctic with the Aircraft Owners' and Pilots' Association (AOPA).*

[2]  *The Chronicler added a note here commenting that this feat appeared to have been done before, and cited Hatton Turnor's 'Astra Castra' as the source of this information (see Chapter 2).*

Reports of the exhibition carried two descriptions of man-powered aircraft, one being a Charles Spencer machine:

"No. 8 - by Charles Spencer, consisted of long wedge shaped aeroplane with an under web or keel set vertically. This combined with two short wings moved by the arms. By a preliminary quick run the Inventor of this arrangement has been enabled to take short flights to the extent of 160 feet. The weight of the body during this time being sustained by the planes." (See also *Fig. 24*).

"No. 9 - by W. Gibson, consisted of two pair of wings worked by the feet and arms together, during which action the shafts of the wings swivelled in their sockets, so as to give a feathering movement similar to that observed during the flight of a bird. The Inventor states that in a former machine, having only one pair of wings of lighter construction, their action of the air during a vigorous down stroke with the feet on the treadles, was sufficient to raise the man and machine; thus showing that a man has sufficient muscular power to sustain his weight during one impulse; but from a faulty arrangement the support could not be continued during the upstroke. But these experiments have so far convinced the Inventor of the practicability of flight by such a means, that he is now engaged in constructing a more perfect machine that will give a continuous supporting action."

During the Crystal Palace exhibition, the Duke of Sutherland offered a prize of £100 to the inventor of a machine which "not being of the nature of a kite or a balloon, shall ascend with a man to the height of 120 feet". Another prize offered at the exhibition, to: "The Exhibitor of the lightest Engine in proportion to its power from whatever source the power may be derived" was won by Stringfellow for his steam engine. This was only the tip of the iceberg as far as the clamour for a small, lightweight engine for aircraft propulsion was concerned, and the development of engines based on steam, electricity, petrol, or some other form of energy, as opposed to muscle power, was receiving an increasing part of the attention of the more serious aeronautical engineers.

In the Annual Report of the Society for 1869, reference was made to a paper by Thomas Moy, an advocate of steam propulsion for aircraft, in which he deliberated on such diverse topics as the relationship between power and wing area, and the flight of the albatross. During the discussions following presentation of the paper, a Dr. Smyth of Ireland disagreed with the speaker on the subject of the relative strength of leg and arm muscles, stating that the former were not necessarily the stronger. He continued: "Muscles are stronger in proportion as they are shorter, and the muscles of the arms, when well developed, are immensely strong".

The Society was often asked to provide financial and/or technical assistance to further the efforts of individual inventors. One such Inventor was Mr. Chalon, who introduced his ideas in a paper given before a meeting in 1870. His design consisted basically of a rigid parachute of approximately 3.6 metres diameter, supporting a cage in which stood the pilot. Two wings, each of about 3 metres span with an area of 3.7 $m^2$, were attached by a swivel joint to the frame around the pilot, so that they could pivot vertically. Strands of rubber connected the underside of each wing with the bottom of the cage. The pilot held two rings, cords from which were looped around the edge of the parachute to the upper surface of the wing. If he then pulled the rings, the wings rose, and on releasing the strings, the rubber strands pulled the wings back down to their original position.

Chalon estimated that the parachute would support approximately 50 per cent of the weight of the pilot when the machine was floating or descending. With a design empty weight of 12.7 kg, Chalon predicted that the craft would successfully fly with an all-up weight of up to 80 kg.  Subsequent references to this machine are conspicuous only by their absence.

The Aeronautical Society naturally became a clearing house for ideas on flight, communications being received from abroad, as well as from members within the British Isles.  One such note, mentioned in the Annual Report of 1871, was received from David Gostling of Bombay, who suggested a man-powered aircraft having wings with the form of those of a bat, and weighing only 9 kg. The propulsive force was to have been obtained by a rowing motion.

In 1873 D.S. Brown[1] of Stoke Newington gave a paper which considered in detail many aspects of flight, aircraft design and manufacture, and means of propulsion.  In the section dealing with the last mentioned matter, he wrote:

"I now come to consider a most important power, namely manual power, and it will greatly curtail what I have to say, if I state in the first instance that in estimating the power required for flight, one of these mistakes so peculiar to the science of aeronautics is almost invariably made.  The one in question arises from supposing that force is necessary to sustain a body in the air, as well as to propel it.

Now force is certainly required for propelling, for it implies motion and a resistance, but, theoretically speaking, no power is necessary to support a body under any circumstances, where no elevation of it takes place.  In practice, however, it may amount to almost anything or nothing according to the conditions observed.

In the case of the toys[2] ... which were sustained in one place by the continuous action of screw propellers, it was enormous.  Small birds and insects, on the other hand, as I explained in a former paper, diminish it to a fraction by intermittent motion.  But with respect to large birds, where the support is merely the effect or the consequence of propelling, no allowance whatever need be made for it.

Now, a fair comparison between the locomotive performance of animals in the air compared with those on the ground will show a result vastly in favour of the former.  It is, therefore, not unreasonable to assume that a man who can propel himself so well upon a velocipede on the ground, will do so still better with a suitable machine in the air.  It so happens that the position in which he can exert the greatest amount of muscular power offers also most resistance to the air, and I need scarcely say that such resistance is very different in flying to what it is in walking.  It would, therefore, be necessary for him to work in a narrow compartment, having its front brought to a very acute angle.  This would diminish the resistance to about a quarter, and two or three working in it, in a line, would reduce the proportion much more."

[1] *Brown took out a number of patents on aircraft, the first being in 1852 (No. 155);  three others were recorded.  All of his claims, however, related to aircraft with steam engines or other non-human power sources.*

[2] *It is uncertain to which toys Brown was alluding.*

Brown showed a keen appreciation of the rudiments of aerodynamic drag, but his arguments concerning the power required to lift a machine vertically and the relative performances of land-based animals and birds are inexplicable. His heart was not in man-powered flight, and he concludes the above paper with remarks in praise of small steam engines as power sources for aircraft.

A paper more directly concerned with man-powered flight was that given by Cradock at the Society's Annual Meeting in 1878. Entitled *The Power Possessed By a Man in Relation to Aeronautics*, it had been written several years earlier. Introducing his paper, Cradock stated that he was particularly optimistic about the successful future of aeronautics.

He opened by likening the power and transmission system of a man-powered aircraft to a man climbing a revolving staircase, the power produced by each step being transmitted to a pinion which gears into a vertical rack. "On this revolving stair, being set in motion by a man placing his weight on successive revolving steps, it will elevate the man to as great an extent as if he employed the labour in walking upon ordinary stairs". Believing that a man possessed sufficient power to enable him to fly, Cradock considered that the effective utilisation of this power was the only step to be overcome. His analogy between an aircraft and a revolving staircase was expanded: "Now we have the substitute for the rack in the air, and the pinion in the wings, which, with proper proportioning of the leverage to the altered circumstances, we get the substitute for the revolving stair, which would also be a failure without the proper proportioning of the leverage".

To the question of 'leverage' Cradock attaches great importance:

"From the nature of the case, long levers in the wings are a necessity, and short levers at all other points are most desirable. By my mode of concentrating all the power in the hand and foot shaft, we are able with small counter-shafts and small wheels to adjust the leverage to the power and the resistance to the greatest perfection, and not yet add above three pounds to the weight, thus the power and leverage questions are placed beyond a dispute, as practically settled; as the only objection to the wheel gear was that the power would be thus increased at the loss of speed. The answer to this objection is, that as the object is to get the substitute for the rack and, as the only assistance to the wings that can at all equal the power, when applied with proper leverage and surface must at least be equal to that the rack had to sustain, it is quite clear that man has the power to put such a force upon the falling wings to produce a resistance underneath them greater than his weight, and when this is done there is no more reasonable doubt about it ascending with the double pair of wings upon this rack of air than upon the rack made of iron; for let it not be forgotten that in both cases the power is the same, leverage is as adjustable in the one case as in the other."

He terminates his paper thus:

"Believing this Society to have truth and useful practice for its object and like myself, to have no desire to be considered as making vain efforts to accomplish what it is not possible to accomplish; but at the earliest moment to test the matter at small cost in such way as shall show our assailants that ignorance in them and not folly in us was the cause of their ridicule of us, I have read this paper, and I am desirous to assist and so far as all past thought, labour and expense, I give it freely; my

aim in these labours never was profit, but since 1859, when they would let
me do nothing else, I have exhausted this subject, and rendered the
mechanism as simple as it is perfect.  I have spent a life on mechanical
and scientific improvements, but I never failed mechanically or scientif-
ically on any one question I took in hand;  perhaps the reason was that I
took good care to see the end from the beginning;  but I certainly failed
to see in time the way in which those for whom I laboured[1] would treat me."

The Chairman of the meeting, James Glaisher, F.R.S.[2], was sceptical in his
closing remarks on the proceedings:

"I was under the impression that man had not sufficient power to raise
himself and all my thoughts and reason and investigation have led me to
that conclusion, but I am very glad that a man with large mechanical
knowledge and experience, who has thought of these things for many years,
has clinched his arguments by the observation that the older he grows, the
more convinced he is of their accuracy.  Assuming that this is a thing that
man can do.  I hope Mr. Cradock will apply his theory and that we may live
to see the application with our own eyes, because seeing is believing.
Then we will thank Mr. Cradock for his invention, as we thank him now for
his paper."

## THEORY INTO PRACTICE — INCLUDING AN EARLY HELICOPTER

While speculation remained the order of the day in London, activity across
the Channel was still at a high level.  Drzewiecki was testing his most complex
'Luftvelociped' which consisted of two lifting rotors, mounted one above the
other, of 4.5 metres span, and also two propellers, of 1.8 metres diameter, one
on each side of the pilot, who turned all four fans with a treadle action.  No
flight was recorded.

In 1871 Prigent designed a tandem-winged ornithopter with which he intended
to simulate the flight motion of the dragonfly.  Danjard in the same year
proposed a machine with a triangular sail fixed onto the front pair of his
tandem wings, and a pentagonal sail fixed to the trailing edge of the rear
pair.  Between the rigid wings were two propulsive flappers, and further
assistance was provided by means of a pusher propeller at the end of the
fuselage.

Vincent de Groof, a Belgian shoemaker, constructed an ornithopter in 1874.
The basic framework of the machine, which had a wing area of about 20 $m^2$,
resembled a coat stand, in which stood the pilot.  De Groof used his arms for
flapping the wings, and a large tail was moved by the pilot's feet.  The
authorities in Belgium and France were loath to permit de Groof to carry out
flying trials in public, and so he came to London and ascended beneath a
balloon from Cremorne Gardens on 29th June, 1874.  No free descent was made
on this date, but on July 9th he ascended, cut the ropes supporting his
machine from the balloon, and crashed to his death, *(see Fig. 20)* having been
unable to control the large flying surfaces.

---

[1] *Cradock spent a considerable part of his capital on improving steam engines,
but his career in this field appears to have ended on a sour note.*

[2] *The Chairman would normally have been the Duke of Argyll, then President of
the Society.  On this instance, however, he was unable to attend.*

In 1878 Dandrieux proposed an ornithopter to imitate the airscrew action
of a bird's wings.  This resulted from the work of Professor J. Bell Pettigrew,
who, in his treatise entitled *Animal Locomotion* realised that the tips of the
wings of birds acted like airscrews.

Helicopters attracted the attention of a number of man-powered aircraft
designers of this period, among them Renoir, who in 1872 constructed such a
machine with a 4.6 metre diameter lifting rotor.  Also Faure and Godard
constructed a single seat man-powered helicopter, with no provision for
forward flight.  The pilot clung to a single leg carrying a set of pedals, and
a chain leading to the rotor.  The only real successes obtained were with
rubber driven models, and the use of rubber motors for model ornithopters was
also widely advocated during this period.

The physician and natural scientist, Hermann von Helmholtz, working in
Germany, concluded on comparison of the musculature of the stork in relation
to its weight, that man was unable to fly by means of his own power.  Helmholtz
compared the power outputs of various birds of different sizes, and related
this to the size of muscles.  His quantitative conclusions were that if a stork
weighing 4 kg required 0.5 kg by weight of muscle to fly, then a creature
weighing as much as a man, typically 68 kg, would need 8.5 kg of muscle, i.e.
12.5 per cent of his total body weight would be muscle.

The mis-interpretation of these results by other workers did not aid the
advances in man-powered flight.  For example, von Mises assumed that use could
be made of the calf and chest muscles only, which weigh a mere three per cent
of a man's body weight.  The total musculature weight of a man is far in excess
of the 12.5 per cent limit laid down by von Helmholtz.

## CARL BUTTENSTEDT

One of the pioneers in the scientific observation of bird flight, Carl
Buttenstedt, who died in 1910, explained the principle of man-powered flight
as he interpreted it with reference to his own work, and a direct translation
of a paper he wrote for the journal *Der Stein der Weisen* in 1890, is of
interest:

"First imagine ... a man supported under two light elastic windmill sails,
constructed like the pinions of bird wings;  the man jumps off a high
cliff like a condor.  Supposing it were calm, then the man moves downwards
and his wings, under which he is hanging, are subjected to a pressure from
below.  It is as though air pressure acts upwards.  The result is that the
tips of these wings bend upwards and then in the same way their inclined
air-components flow rearwards;  because the falling motion is a gentle
beating motion.  A horizontal stress is therefore set up in the flying
surfaces because the heavy body of the man cannot be pulled out of the
inertia of its vertical fall as quickly as the wing tips would like to do.
If the man is heavier than a normal man he will fall faster, and the air
pressure will in turn become greater so that the tension consequently also
increases.  The tension force is an equivalent effect to sail force, hence
a heavier body will sail more rapidly than a light body under identical
conditions.  With such wings a man will never fall vertically, but will
always be driven forward like a sailing boat, and in the case of a bird
this action is developed to such an extent that it glides almost horizontally, without flapping its wings.   ... The man must be attached under the

wings by means of an elastic suspension system in such a way that by means of handles he can rotate the wing surfaces longitudinally to enable him to produce at will the required angle of flight.

As an aid to soaring he should first of all fasten a pair of small flapping control surfaces to the lower part of his legs and use these for upward, downward and lateral control, and also for driving himself forwards. The latter action is produced by spreading his legs as far apart as possible, twisting himself at the hips and striking upwards with the right leg, at the same time moving the left leg downwards.

The simpler the complete device, the easier it will be to handle it, and more surely will it achieve its purpose. The device on the legs acts like the wing movements with which Hargrave propelled his machine, but it also resembles a screw, with which one can push forward and with which one can also slow down. By beating more strongly with the right or left foot one can turn to the right or to the left; by lowering both surfaces one changes direction downwards and by raising these surfaces the direction of flight is directed upwards. By turning one leg about its longitudinal axis one can produce a simple ships rudder and by this means direct the craft sideways and thus, when the wind is strong, create a rudder with both flapping wings, which serves to maintain horizontal and vertical direction. There is no reason to believe, however, that greater leg forces are required to keep stretched out in this way; on the contrary, the surfaces are constantly forced into the windstream by gliding forces of the craft and the legs will therefore be supported more by the surfaces, instead of the legs having to support the surfaces. Only when something has been achieved along these lines can one go on to make the front wings to a certain degree movable, both upwards and also forwards and rearwards, in order to be able to balance and soar more safely in irregular winds (see *Fig. 21*).

One can think of further improvements and devices to make it possible to fly from open country as well. The attempt must be started from elevated points, then be made over water or woodlands to soften the effects of a possible crash.

To consider once again the mystery of bird flight, my studies of nature show this to be nothing more than a conversion of the force of gravity into a gliding force, or a conversion of the work done in falling into soaring work. The only part of this energy which is lost during conversion is that which is expended in friction with the air. If we replace this small energy-loss due to friction with the medium by human power, then we obtain, in accordance with the laws of the conservation of energy, the force of gravity which has been converted into soaring work for the duration of the longest flight. In the weight of his body the flyer has therefore brought with him the most powerful motor for his flight and only needs to maintain this (his flight) by means of his own power so that he is certainly in a position to keep himself in the air as long as a cyclist on the wheel. But there can certainly be no question that during this time he will cover a very different distance from that covered by the cyclist."

Whether Buttenstedt persuaded colleagues to 'jump off high cliffs like condors' is, unfortunately, not recorded.

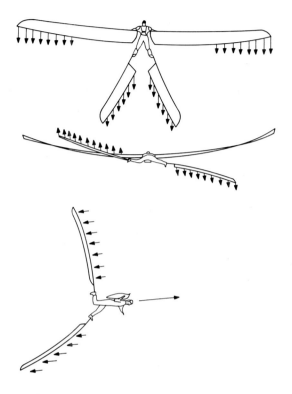

*Fig. 21*   Buttenstedt's proposals for controlled flight in 1880

## ECLIPSE BY THE INTERNAL COMBUSTION ENGINE

Karl Steffen, an expert of his period on the theory of flight, designed a 'wind flying machine', the principal features of which are described below. Steffen stated that the aim of his design was to enable a man to undertake free flight by means of his own muscular power and skill.  The machine had a pair of flapping wings mounted on a frame made primarily of aluminium and each was covered with stretched silk.  The drive for each wing was applied by a foot lever, transmitted by arms to the wings.  The total weight of the machine was to be 10 kg, with a wing area of $7m^2$ and a span of 7 metres.  To some extent the machine designed by Steffen seems to have resembled the early ornithopter designs of Leonardo da Vinci, but it is not known how far Steffen proceeded with the construction and testing of his aircraft.

With the proposals of Buttenstedt and Steffen at the end of the nineteenth century it was felt by many that the era of flapping wing aircraft had come to an end.  The enthusiasts for man-powered flight became overshadowed by the inevitable trend towards adoption of the internal combustion engine, and this was aided by the model makers who, even using rubber powered ornithopters, could make a successful flying machine.  Lack of a suitable lightweight engine still daunted the pioneers, but steam was available, and although not success-fully used, did take precedence over muscle power with aircraft of the period requiring powers of the order of 15 kW (20 hp) to fly[1].

With the advent of fixed wing aircraft powered by the internal combustion engine, an incentive to rethink the philosophy of man-powered flight was presented, although this did not really emerge in the form of hardware until the early 1930's.  Before then there was still to be a 'silly season'.

## THE UNITED STATES — 'BEFORE THE WRIGHT BROTHERS'

European aeronautical circles seemed virtually unaware of work being carried out in the United States of America prior to the flights of the Wright Brothers.  However, details of a few attempts at man-powered flight on the other side of the Atlantic are on record.

In 1883 John Montgomery constructed a monoplane glider which was unfortun-ately destroyed in a crash.  Later, in 1886, he made a machine with bird-form wings which could rotate about the spanwise axis.  This aircraft failed to fly. William Cornelius was another experimenter of this decade, but his ornithopter, based on similar operating principles to de Groof's machine, also came to nothing.

A patent was taken out on a flying machine by Reuben Jasper Spalding in the United States in 1889.  Spalding, who lived in Rosita, Columbia, had worked on his device for a number of years prior to obtaining Patent No. 398984, in which several claims were made concerning the "latent means of high speed transport" designed to enable him to "travel along all routes in every possible direction at the greatest possible speed and in the most direct line".

The basic form of Spalding's machine, shown in *Fig. 23,* consisted of a balloon from which the pilot was suspended.  In order to provide additional lift and forward motion, as well as the ability to steer, a pair of wings and tail were strapped to the pilot's back using a corset.  The wing span was about 3 metres, and the tail resembled that of a bird, artificial feathers also being used on the trailing edge of the wings.  Actuation of the wings was by cords attached to the pilot's wrists, and a lever mechanism was used to adjust the tail position.  The inventor did not consider the balloon absolutely necessary, claiming that the wings would operate as efficiently as those of an eagle.  No doubt Spalding failed to appreciate that his weight was considerably greater than that of an eagle, and that wing efficiency would be very low.  In the patent, Spalding made further suggestions for improving his design, including the use of electric power for flapping the wings.  Although a model of his aircraft was constructed, there is no record of a full scale prototype.

---

[1] *One horsepower is approximately 0.75 kW.  A good cyclist will therefore produce about 0.30 kW over an extended period.  (See Appendix II for Conversion Tables).*

*Fig. 23.* The balloon-aided ornithopter proposed
by Reuben Jasper Spalding in the USA, 1873.

## 'A PARADOXICAL DESIGN'

As the end of the nineteenth century approached, the activities of the
Aeronautical Society were becoming more concerned with fixed wing aircraft
and their propulsion by means other than man-power.  However, the Journal of
the Society, published at that time every three months (now monthly), still
contained a number of contributions on bird flight and its relevance to man-
powered efforts.

A review paper entitled *Flight and Flying Machines - Recent Progress*,
written by J.D. Fullerton, a Major in the Royal Engineers, and published in the
July, 1897 edition, did not allude directly to man-power, but gave some
considerable insight into the contemporary thinking on the form of bird flight.

Fullerton accumulated a considerable amount of data on the performances of
birds and, in his own words:

"... a great deal of information as to size, dimensions, weight etc. is
available, owing to the researches of Mouillard, Chanute and others, and
the following details are of interest, as they give some sort of idea of
the probable dimensions of the flying machines of the future.

Sustaining surface - the weight carried per square foot of wing surface[1]
of some of the principal birds, is as follows:

|        |           |
|--------|-----------|
| Swallow | 0.276 lbs |
| Gull    | 0.435 lbs |
| Rook    | 0.575 lbs |
| Duck    | 1.158 lbs |

---

[1] *Commonly referred to as 'Wing loading'.*

He stated that the 'motor' of a bird, the pectoral muscles, weighed from one quarter to one sixth of the total weight of the bird.  The power outputs cited by Fullerton, based on Penaud's data, are of interest:

|         |                          |
|---------|--------------------------|
| Peacock | 1 horse power per 66 lbs |
| Pigeon  | 1 horse power per 57 lbs |
| Sparrow | 1 horse power per 48.5 lbs |

In this context it is interesting to note the maxim of the day that "a flying machine was required to develop about one horse power for every 50 lbs of the total weight of the machine when fully loaded".  The results of Penaud's bird studies are particularly close to this;  "50 lbs per horse power" figure, but such figures are well beyond the capabilities of a man.

The article made several references to the various flight manoeuvres; "starting, turning, gliding, soaring, hovering and alighting".  In referring to power units, the author stated that most existing motors were too heavy and unsuitable for aircraft, but he did not say that man-power was acceptable! Under a sub-section dealing with *Applications for the Flying Machine in Peace*, Fullerton mentioned aerial yachting as a sport which could possibly compete with marine yachting in popularity;  however, he could be envisaging gliders or man-powered aircraft here - more likely the former.

In the October issue of the same year, another Major of the Royal Engineers, R.F. Moore, reported on some experiments he had carried out to determine the 'power and means necessary for flight with wings'.

He made a model of a flying fox, although it is uncertain what purpose this served.  Of more use was his clockwork ornithopter, which he suspended from wires.  During operation he measured the stresses occurring in the wings.  His experiments led him to conclude that a man-carrying machine would ideally have four wings and a motor, the power requirement being in excess of that capable of being generated by a man[1].  (An earlier system is shown in *Fig. 24*).

Also in this edition was a report on a visit by Society representatives to witness an experiment:

"We recently had an opportunity of witnessing a trial of this invention. (Waeldes' Aerial Propeller).  The apparatus consists of what may be described as a pair of feathering paddle-wheels, though each wheel is fitted with only three floats, these consisting of a circular frame three and a half feet in diameter, covered with cloth, pivoted on trunions.

These paddles are actuated by man-power by means of endless chains, two men working pedals similar to those of a bicycle.  The angles at which the plates strike the air are controlled also by endless chains, so that by means of a pair of levers the propellers can be made to thrust up or down, forwards or backwards.  The present apparatus, which is but a large working model, is neatly constructed of steel tubing, weighing about 115 lbs.

It is suspended from one end of a beam, the other end bearing a counterpoise. Its exact power has yet to be determined."

---

[1] *Moore seems to have meant this machine to be an ornithopter.  He suggests the imitation of pectoral muscle function by using springs.*

This invention was the subject of Patent No. 9108, dated 30th April, 1896. No further reference to this device was found, and it is likely that Waeldes discontinued work following unsuccessful tests using his counter-balance system.

Under the heading 'A Paradoxical Design', the Aeronautical Society commented at length upon a proposal for a man-powered aircraft put forward by Mr. Alex Adams, a civil engineer of Sydney, Australia. In a paper sent to the Society, Adams contends that "if a machine be constructed with beating wings, its flight may be indefinitely prolonged". This statement is not taken out of context, but more reasonably, Adams continues: "... man, however, does not possess anything like sufficient muscular power to operate loaded wings of eighty or ninety feet area". The apparatus described is worked "without any exertion on the part of the occupant" and yet "no steam, electric, or other engine is required".

The Society kindly gave the author the benefit of the doubt in stating that "we must own that we fail to understand the principle from the description given ..."

W.C. Congreve, the nineteenth century inventor more noted for his pioneering work in the field of rocketry, was the designer of the first recorded 'Cyclogiro'[1], propelled by man-power. Although never constructed, a description of the machine was published in *Mechanics Magazine* of 22nd March, 1828. The article concluded with rather fanciful predictions concerning further prototypes capable of flying up to 1900 km in one day and requiring over 2 kW power.

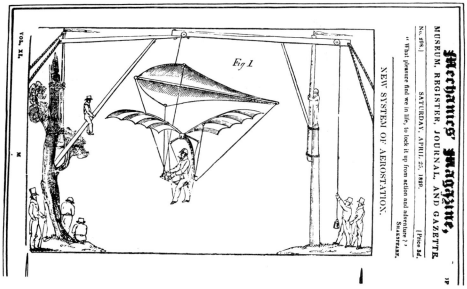

*Fig. 24*  A 'Manumotive Flying Machine' proposed by Charles Spencer in 1868.

---

[1] *Cyclogiros derive lift and propulsion from blades mounted at the ends of rotating arms. The blades in turn are able to pivot about their own axes, in order, for example, to maintain a particular angle of incidence. The overall effect resembles a paddle-wheel.*

The nineteenth century ended quietly on the aviation front in Britain.
The period of great initiative had passed over to the Continent by the 1870s,
particularly to France and Germany, and in Chapter 4 a few steps back in time
will be taken by way of introduction to cover this.

It is appropriate that a paper entitled *The Flapping Flight of Aeroplanes*,
given by Maurice F. Fitz-Gerald, Professor of Engineering at Queen's College,
Belfast, to the Royal Society, should be reproduced in the Aeronautical
Journal of July, 1899.  The author concludes:  "All that can be inferred,
therefore, is that provided that a sufficiently high speed of flapping (of
wings) is attainable, we may reasonably anticipate that the horse power for
hovering need not differ very materially from that for progressive flight".
Using the current vernacular, this paper was undoubtedly the 'non-event' of
the aeronautical scene, and was indicative of the lack of progress made in the
field of flight, man-powered or otherwise, during the previous three decades.

* * *

*Fig. 15.*   Fanciful representation of an early aerial
             Casanova.

*Fig. 13.* The reverse of a silver disc
showing an ornithopter type
proposed in 1799 by Sir George
Cayley.

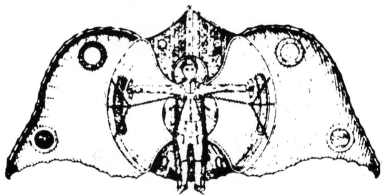

*Fig. 16.* Bréant's design of 1854, in which elastic
was used to return the wings to the top of
the stroke.

*Fig. 17.* The Le Bris ornithopter.  Depression of the
wings was by the pilot's arms, but a spring
return was incorporated.

*Fig. 18.* 'Aerial man passing Greenwich Observatory!' A prediction of the year 1843.

*Fig. 19.* Crystal Palace, July 1868.  The first Aeronautical Society Exhibition.  Steam power,
balloons, and muscle power were all represented.

*Fig. 20.* Portrayal of the death of Vincent de Groof
in 1874 during an attempted ornithopter
flight. He fell into a Chelsea Street.

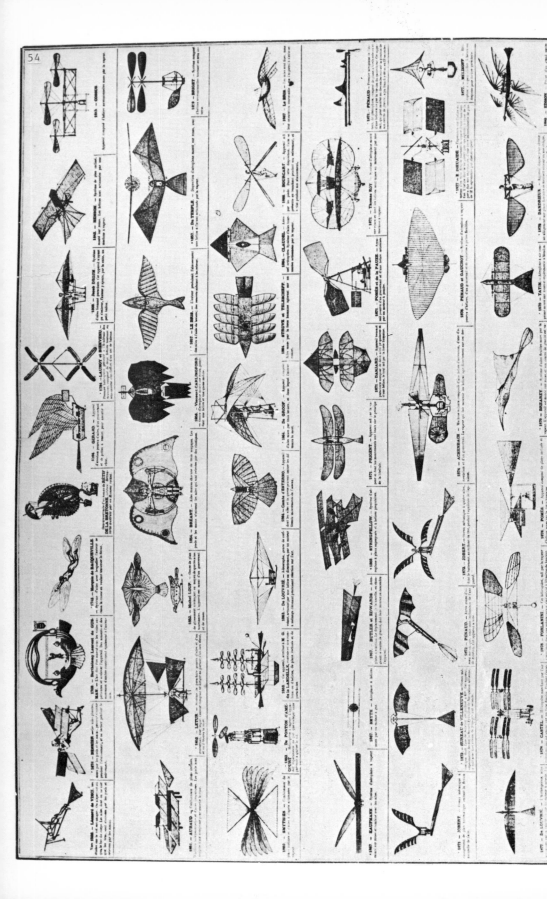

54

E. DIEUAIDE, 18, Rue de la Banque. — PARIS.

*Bureau pour les* BREVETS, DESSINS, CONSTRUCTION D'APPAREILS

*4*

# The Sublime and the Ridiculous

*4*

Problems of communication have continuously plagued scientific development, and it is only recently that an appreciation of the real importance of good communication is showing itself, although it is still not sufficiently wide-spread. To some extent, therefore, it is appropriate to record the research of Otto Lilienthal in the chapter dealing with the early twentieth century, rather than in Chapter 3, as his work had little effect on aviation, or the aspects of it concerned with man-powered flight, until the turn of the century.

## OTTO LILIENTHAL

Lilienthal, a German, was born in 1848 at Anklam, in Pomerania, and received his technical education at Potsdam and Berlin. He became interested in man-powered flight at a very early age, and was assisted by his brother Gustav in early experiments with ornithopters. Although much of his development work was carried out on fixed wing gliders, and it is for this work that he is most well remembered, his enthusiasm for the ornithopter concept was to be particularly long lasting, and in 1889 he wrote: "Natural bird flight utilises the propertie of the air in such a perfect manner, and contains such valuable mechanical features, that any departure from these advantages is equivalent to giving up every practical method of flight".

Lilienthal studied bird flight in considerable depth, and was one of the first to appreciate that bird propulsion was not obtained solely by flapping, but with the aid of propeller type movements of the outer wing feathers. His studies on bird aerodynamics were published in 1889 in a book entitled *Der Vogelflug als Grundlage der Fliegekunst* (Bird Flight as the Basis of Aviation). Writing of the Lilienthal brothers' efforts to translate the results of studies on bird flight into a practical flying machine, Gustav, in the preface to this book, described the construction and tests of such a machine, probably built by the brothers in the late 1860s.

"He (Otto) brought with him a bundle of palisander sticks ... to work the hard palisander wood was no small matter; we pointed and rounded the sticks which served as quills for two wings, 3 metres long each. The feathers of these quills were represented by a series of large goose feathers which were sewn on tape.

For this purpose we had purchased all the feathers which were obtainable in our town, and this is no mean accomplishment in any Pomeranian town."

(This was the second machine built by Lilienthal, the first being a rather crude set of wings which were strapped to the arms. At this time the brothers were still at school, and practiced wing-flapping exercises at night to avoid being mocked by their school colleagues.)

"The wings were fastened to two hoops, one of which was strapped round the chest, and the other round the hips, and by means of an angle lever and stirrup arrangement of the ropes we were enabled to beat the wings up and down by pushing out our legs. The single feathers were arranged to open and close on the up-and-down stroke, and the arrangement worked perfectly ... We did not heed the lesson taught by our storks[1], but suspended our apparatus from the beams of the roof and began to move the wings. The very first movement of our legs brought about a jumping at the suspension rope, and as our position was nearly horizontal, we were most uncomfortable. When drawing up our legs, that is, when the wings moved upwards, the whole arrangement dropped down and hung on the taut rope. The lifting effect due to the beating down of the wings amounted to 20 cm. This was at least some success ..."

It appears from his writings that Otto Lilienthal underwent a transition period around 1890, when he commenced work on gliders. It is noticeable that his lessening interest in flapping flight was also based on observations of the birds, and his yearning to "soar upward and to glide, free as a bird ..." However, although his gliders, one being illustrated in *Fig. 25*, which he successfully flew until his death in August 1896, had fixed wings, he was still aware of the need to generate power to take off and climb to an altitude at which gliding could safely begin. Apart from a brief period in 1893 when he unsuccessfully attempted to make flights in a machine powered by a 1.5 kW motor, he seemed to maintain a faith in the ultimate success of a man-powered aircraft. His conclusions in his treatise on bird flight uphold this: "Motive power and force are numerically limited, but not so skill. With 'force' we are ... confronted by permanent impossibilities, but the progress of our skill can only be temporarily checked by difficulties."

## BIZARRE RUSSIAN EXPERIMENTS

About the same time as Lilienthal was working on bird flight, a number of Russian scientists were carrying out similar studies, some of which ended with the construction and attempted flights of man-powered aircraft. Vinogradov, in a report[2] published in Moscow in 1951, briefly outlined the history of these attempts, and commented upon the work of various Russian academics of the period.

It is evident that many of the early Russian experiments were based on the belief that the ornithopter was the most likely aircraft to succeed, with power provided by muscle or machine. (The Russian inventors were, of course, not alone in this presumption). Professor N. Zhukovskii, in a report entitled *On Aeronautics*, published in 1898, expounded upon the theory of the ornithopter, and his maxim for optimum performance was: "A perfect flying machine driving down the maximum quantity of air at the minimum velocity".

Vinogradov also makes reference to the earlier work of V. Mikhnevich, who wrote a thesis in 1871 on the relationship between bird flight and its applic-

---

[1] *Lilienthal spent his formative years in close observation of the flying action of birds in general, but the stork's wing motion impressed him most.*

[2] *I.N. Vinogradov. 'Aerodinamika Ptits-Paritelei'. Moscow, DOSARM, 1951 (Available as RAE Library translation 846, January 1960.)*

ation in an aircraft that he designed.  In his flapping wing machine the
wings were connected by hinges to a cross beam, and levers were attached to
them, brought together by a spring.

    In 1880 Lieutenant V.L. Spitsyn designed an apparatus in which four flapping
wings rose edge-wise, to reduce drag, and were lowered flat, providing lift, at
a frequency of two beats per second.

    Some of the experiments carried out by Russian ornithologists and aviators
in attempts to study the flight of birds as applicable to man-powered flight
bordered on the bizarre.  Dr Arendt, in 1890, froze living birds with their
wings and tail in various positions, sent them up to an altitude of 50 metres
on a kite, and then dropped them, observing their subsequent gliding, and even
soaring flight.  Almost twenty years later, in 1908, the aviator Shiukov
performed a similar series of tests on crows at Tiflis.  He found that when the
large feathers of the trailing edge of a bird's wing were glued together, it
completely lost the ability to fly, whilst joining those in the middle part of
the wing compelled it to use a form of rocketing flight, with more frequent
wing beats.  The bird soon became exhausted in this flying mode.

    Shiukov, however, must have received some encouragement from his experiments
for he proceeded in the same year to construct and test a man-powered ornithopter.  The few details available suggest that only the outer wing sections
moved.  Flight testing took place at Tiflis, and the site, Makhal Hill, would
lead one to believe that downhill runs were used in an attempt to take off,
although no report of its success or failure was found.  The aircraft was a
biplane with the ornithopter sections attached to the upper wing, and the
pilot used his feet to provide power for moving the wing sections.

    At the time Shiukov was glueing crows' feathers, another Russian scientist,
Dr E.P. Smirnov, was carrying out almost identical experiments on pigeons,
which he described most vividly in his reports on the work:

    "Taking a fine lady's silk veil, I glued the finest cigarette paper on both
    sides of it.  The result was an exceptionally light fabric which ... did
    not easily tear when subjected to pressure.  I then acquired from a
    pigeon-fancier I knew at Shchukino, a good tumbler-pigeon which could soar.
    Having anaesthetised it with chloroform, I spread it out on its back on
    the table and stretching out its wings, stuck the tissue over the underside
    in such a way that the wings could freely extend and fold like a Japanese
    folding fan.  It will be readily understood that in this way the gaps
    formed between the primaries (main wing feathers) when the wing was spread
    out were no longer open but covered by the light tissue.

    When this troublesome and somewhat ticklish operation was completed, I
    brought the pigeon round and after it had sufficiently rested and recovered
    in a narrow cage, I released it outside to see how it would behave.

    Realising that it was free, the tumbler made a few strong beats with its
    wings and rose up into the air.  But its flight, as I had anticipated,
    rather resembled the characteristic fluttering or even, if one may use the
    expression, flittering of a cabbage butterfly ..."

    Smirnov described how the pigeon soon became exhausted, and also noted an
abortive attempt at soaring.  His conclusion was that the ability of a bird to
soar depended to a large extent on the presence of slots between the feathers,

i.e., a 'penetrability', and that designers of ornithopters must reproduce these slots if they were to approach success.

## KEITH-WEISS AND OTHERS

Activity of a more experimental form than that prior to the turn of the century was predominant in Britain between 1900 and 1925.

It is worth recording the work of E.P. Frost, who, after a decade of study of bird flight, constructed a number of powered ornithopters. Two of his machines are the earliest of their type to survive to the present day. The larger, weighing 291 kg, was powered by a steam engine and is preserved by the Shuttleworth Trust. Made in 1900, the wings were constructed of cane and silk with hundreds of natural feathers attached to resemble a bird's wing. The smaller, made in 1903, had a 4.88 metre span. Power was by a single cylinder B.A.T. petrol engine of 2.3 kW. Neither machine flew, and the latter is now at the Science Museum, London, and is illustrated in *Fig. 26*.

S.R. Hewitt constructed an ornithopter in 1908, the wings consisting of planes of 9.68 metres span with a chord of 1.51 metres, and an additional flexing portion 0.61 metres wide at the wing trailing edge. The machine had a tricycle undercarriage and fore and aft control surfaces.

The next year saw the manufacture of two similar machines, in that both were, as man-powered aircraft, intended only as an interim solution to the problem of lack of availability of a suitable engine. One of these was the Druiff-Neate 'cyclo-aeroplane', constructed by C.G. Spencer & Sons of Highbury, London. The 'cyclo-aeroplane' was a single seat monoplane with a 1.21 metre diameter pusher propeller driven via foot pedals. With a span of 6.10 metres and a wing area of 9.29 m², the aircraft had an empty weight of only 22.7 kg. Although the manufacturers intended to fit an engine at a later date, it is unlikely that this conversion was carried out, and as a man-powered aircraft it was unsuccessful.[1]

The second aircraft was also a single seat pusher propeller design, a biplane, designed by C.H. Parkes. The airframe was mounted on a bicycle undercarriage. A biplane front elevator was fitted. Reports indicate that the machine left the ground for a metre or two when running down a slope. The maximum speed reached on level ground was 14.5 km per hour. C.H. Parkes built a second biplane later in the year - this time he installed a 3 kW engine.

In 1912, as a result of co-operation between Alexander Keith, an elderly Scotsman with a medical background who had made a special study of the behaviour of muscles in a bird's wing, and José Weiss, the Keith-Weiss 'aviette' was born. This was an ornithopter in which foot pedals were used via a lever system to flap the wings with a beat of 0.9 metres. Initial trials were performed with the wings supported in slings between two trees, and the machine reached the stage of free flight trials under ballast following launches from Amberley Mount, Sussex, (see *Figs. 27 and 28*).

The 'vital statistics' were: Span 7.02 metres; length 5.8 metres; wing area 7.44 m²; empty weight 43.1 kg; AUW 104.5 kg; speed (estimated) 48 km/h.

[1] *Peter Lewis 'British Aircraft, 1809 - 1914'. Putnam, London. 1962.*

Less information is available concerning the 'Wood Glider', of the same
year.  Constructed by B. Graham Wood, the aircraft was a small unorthodox
triplane ornithopter with a pronounced wing stagger.  The Rickman tandem
tricycle helicopter, which appeared in Jane's of 1913, was built in the same
manner.  A multi-bladed rotor of approximately 4.5 metres diameter was rotated
by the two cyclists via a shaft and bevel gear.  The machine was constructed
in 1908.

Reference was made to another man-powered aircraft of the period in a
rather unusual context, an advertisement in *The Aeroplane* of 26th October, 1911:

"CYCLEAEROPLANE, constructed by German workman in Lancashire, illustrated
and described in Cycling, 1st June, 1910.  Death cause of selling:
what offers?
                        Tom Wilding, Standish, Wigan."

## F.E. PALMER

Although the number of attempts at man-powered flight in the British Isles
during this period was comparatively few, the work of F.E. Palmer is worth
recording as an exercise in sophisticated design (compared with some European
contemporaries) and perseverence.

Palmer, an ex-member of the Royal Flying Corps, gained his knowledge of
aerodynamics and the theory of flight during his service career.  The first
man-powered aircraft made by him, denoted 'No. I', was a biplane with a span
of 4.27 metres and an empty weight of approximately 23 kg.  This machine was
primarily constructed to prove the structural design, but incorporated all
control surfaces and a large wooden tractor propeller mounted at the front of
the fuselage.  Construction took place in 1919.

The second machine, 'No. II', utilised many of the components of 'No. I',
including the wings and undercarriage.  A slimmer fuselage was constructed on
which to mount all these units.  A 1.53 metre diameter propeller was used,
being manufactured using three laminations of deal.  The maximum chord of the
twin blades was 0.153 metres, and the minimum thickness was of the order of
12 mm.  The pilot used his legs to provide power, the gear used being adapted
from a Lloyd chainless bicycle, which used a bevel drive from the pedal wheel
rather than the conventional chain.  A large chain wheel was fixed to the shaft
leading from the bevel gear, and a chain drive to a free-wheel pulley on the
propeller shaft completed the transmission system.  The total ratio was 11:1,
giving a propeller speed of approximately 600 rev/min for a comparatively
sedate cycling speed.  Tests on the aircraft carried out during 1920 and 1921
showed that although a speed of approximately 27 km/h could be reached using
propeller thrust alone, insufficient power was available for take-off.

'No. III' machine had an upper wing of considerably greater span, in an
attempt to obtain additional lift.  Before any serious take-offs could be
attempted, the machine was blown onto one wing and written-off while being
prepared for tests on Cleeve Hill in the west of England.

The most successful machine built by Palmer was his final one.  'No. IV'
'Magpie', was a high wing monoplane, and utilised components of the previous
aircraft to a considerable extent.  The fuselage of 'No. II' was mated to the
rebuilt upper wings of 'No. III', and a new tail plane and rudder fitted.  The

wing span now totalled 10 metres, with a root chord of 1.525 metres, giving an aspect ratio of slightly over 8. The pilot had a small joystick with which he could operate the ailerons and all-moving tailplane. The aircraft weighed 39 kg empty. An identical transmission system, based on the Lloyd chainless drive incorporated in 'No. II' was used.

Trials lasted almost one week, again taking place at Cleeve Hill, the location shown in *Fig. 29*. After various mechanical adjustments, a take-off was attempted into a steady wind, and with the aid of a troop of boy scouts from Winchcombe, the aircraft succeeded in leaving the ground and, according to witnesses, continued to fly for an estimated ninety metres plus, at an altitude of about three metres. Unfortunately the landing ended in rough ground and the machine nosed over and was badly damaged. No further flights were attempted and the aircraft became part of a local rubbish tip sometime during 1940.

## AIRCRAFT IN CONTINENTAL EUROPE AND THE USA

In Europe and the United States, the years up to 1912 saw a few designs turned into hardware.

R. Schelies of Hamburg in 1906, constructed an aircraft around a bicycle frame, supported by vertical extensions mounted on the front and rear wheel hubs. A cross member between these supports carried the wing which had a span of almost 6 metres. This wing appeared to be divided into three separate sections, each connecting at the wing root. A simple flapping mechanism was provided. Additional lift was to have been produced by a small fixed aerofoil mounted above the main wing, rigidity being obtained using very crude wooden frameworks.

A further attempt to build a man-powered aircraft using a bicycle frame as the main structural member was that of Demcke and Thormann. Their aircraft resembled a box kite on wheels, with two small delta-shaped wings. A two-bladed tractor propeller of 1.525 metres diameter was used, driven by a bevel gear transmission system. Conventional handlebars were used for steering while the machine was earth-bound (which proved to be its only state). No allowance was made for controlling the machine should it have become airborne.

An unidentified aircraft, constructed by Graf Puiseux, shown in *Fig. 31*, which was dated 1912, resembled a Bleriot monoplane, one of the most successful powered aircraft of the period. One of his earlier designs is shown in *Fig. 32*. The wing span was 7 metres and the overall length of the aircraft, 5.5 metres. Empty weight was 33 kg, quite an achievement for a machine of this size, considering the structural materials available. A well-modelled propeller was driven by the pilot through pedals, the primary drive being via a chain, thence to a bevel gear connected to the propeller shaft. Rudder and tailplane motion was to have been used to control the aircraft in flight, and tail wheel steering was used when doing ground runs, for which handlebars were provided.

In the same year Rickman, in the United States, constructed a man-powered helicopter based on a tandem bicycle, and a Frenchman named Collomb unsuccessfully tried to fly an ornithopter.

Soltau, an Austrian, also constructed an ornithopter in the year 1910. He was greatly impressed by the work of the Belgian, de la Hault of Brussels, who

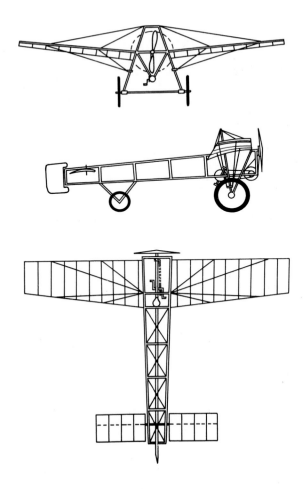

*Fig. 31*  The Graf Puiseux machine with a wing span of 7 metres,
designed in 1912.


built an ornithopter in 1906, followed by a tandem-winged machine with an
additional fixed plane fore of these.  In 1911 he changed this to a fixed wing
biplane.  All of these aircraft are thought to have had motors, but none flew.

In 1909 H. la V. Twining, then President of the Aero Club of California,
constructed an ornithopter based on a tricycle landing gear supporting a simple
tubular framework.  The wing area of 6.04 m$^2$, wing span being 7.92 metres, was
claimed by the inventor to generate 450 kg of lift, but the aircraft remained
earth-bound.

Lehman Weil of New York described an aircraft which utilised the 'breast
stroke' motion for forward propulsion, in US Patent No. 1,569,794.  The power
was supplied by pedal action of the operator, and hinged surfaces, two per

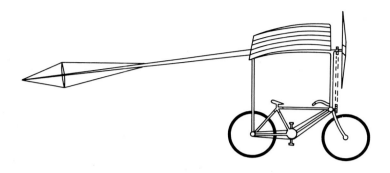

*Fig. 32*   The earliest Graf Puiseux 'Aviette', dated 1909.

side, produced the thrust.  The aircraft was constructed in 1926, and the photograph of this machine (*Fig. 33*) was taken in 1927.

Also in the United States, two years later, in 1928, George White made and tested in secret his 'Orthicopter', at St. Augustine, Florida.  This craft was also pedal-driven, and exhibited a 90° flapping angle, the wings pointing vertically upwards at the start of the lifting beat.  The wings were constructed to simulate those of a bird, and were composed of a number of artificial feathers:  wing span was 9 metres.  During trials, the orthicopter was launched from a speeding car.  It was claimed that fourteen flights were made, the aircraft travelling distances of up to 1200 metres.  However, in a public demonstration it failed to take to the air.

W.F. Gerhardt, a contemporary of White, constructed a man-powered multiplane illustrated in *Fig. 34*.  It is believed that this aircraft never flew.

The Hafner machine, constructed in 1909, had a flapping tail, operated by means of pedals.  The craft was 7.98 metres long, with a span of 7 metres, and weighed 24 kg.  Short hops were made with the apparatus, after which it was abandoned.

Another advocate of the ornithopter, Dr Joseph Cousin of Avignon, France, spent a number of years studying bird flight.  Numerous monographs on this subject were written by Cousin, and in 1911 a machine was built.  The craft was incapable of flight.

The Irvine 'Aerocycloide', a cyclogiro similar to the Congreve machine mentioned in Chapter 3, used disc-shaped aerofoils, four of which were mounted in between two large wheels.  It was constructed but did not fly.

The rotor developed by a Frenchman, Alfred Eugene Pichou, was invented in 1883, but it was not until 1902 that tests were made with this system.  In his proposed 'Auto Aerienne', four two-blade rotors were used to lift and propel the machine.  The system was the subject of French Patent No. 326,476. Pedalling was used to drive the rotors, and a thrust no greater than 15 kg was

recorded in tests.  (This thrust had to provide lift as well as overcoming
drag).  Two years later A. Schmutz constructed another French 'Aviette',
illustrated in *Fig. 35*.  Reference to the photograph will obviate the need for
further comment!

## THE PEUGEOT PRIX

There have been a number of competitions, both national and international,
aimed at encouraging the development of man-powered flight, of which the present
Kremer Competition is probably the most widely publicised, and undoubtedly the
most arduous.

The first major incentive of this kind for man-powered aircraft, one in
which all participants were required to compete in one place on the same day,
making an excellent spectator event, was the Peugeot Competition.

Peugeot, currently one of the major French motor manufacturers, were the
main sponsors of this competition, and continued to support the endeavour for
no less than nine years.  On February 1st, 1912, Peugeot donated 10 000 francs
for an unassisted man-powered flight over a specified short distance.

The purpose of the competition, as stated in the rules, was "... to find
out whether a man is able, solely by his muscle power, to set into motion a
flying machine in such a manner that the machine rises from the ground for a
certain distance and flies".  The following guidelines were incorporated in the
rules:

1.  The choice of the flying machine was to be left to the contestant.  It was
    forbidden, however, to use additional helpers to generate high initial
    velocities to aid take-off of the machine during the start up run.

2.  The first test was to have involved covering a flying distance of ten
    metres, marked by two parallel lines.  The flight was to be carried out
    according to the following conditions:

    (i)  No part of the machine was to touch the ground between the
    two parallel lines during the flight.

    (ii)  Only after the flight had been completed could the machine
    touch the ground again.

3.  In order to be able to prove the practical value incontestably, and in
    order to show that no benefit from a following wind had been obtained, the
    pilot was to fly over the distance after the first test in a reverse
    direction, according to the above conditions.

4.  The person who won the Peugeot prize was to be the one who satisfied the
    requirmments of this double test, beginning from the 1st June, 1912, the
    day of the opening of the competition.

5.  Were the Peugeot prize not won within one year, the donors reserved the
    right to change the conditions of the competition.

The supervisory body for the competition, the Commission de l'Aviette,
consisted of Jacques Balsan, de Knyff, Etienne Giraud, Isaac Koechlin, Rodolphe
Koechlin, Tissandier, Paul Rousseau, Frantz Reichel and Alibert.  This

information, gleaned from a contemporary edition of *Scientific American*, was
given in an article entitled "The Failure of the 'Aviettes'". The author was
singularly sceptical of the attempts of Vincent, Piat (whose aircraft is shown
in *Fig. 36*) et al, at the Parc de Princes:

"... The ignorance displayed by many of the designers of the machines
entered is simply amazing ... Obviously impossible as it was to rise from
the ground even with a monoplane aviette, some of the contestants were
foolish enough to attempt the feat with biplanes."

A premature attempt was made by the French racing cyclist Lavalade on the
23rd May, 1912, at Juvisy airfield. Lavalade's machine was a modified
bicycle with a parasol wing and tail plane. Lift was to be generated by
accelerating up to full speed along the ground, but no reference was made to
the anticipated performance once the cycle left the ground. His flight, more
accurately called an air jump, was observed by sporting authorities. Using a
small spring board, Lavalade was able to clear a ribbon at heights of 10 and
20 cm. The greatest distance jumped was 1.10 metre. (No doubt long jump
champions of the day could achieve six or seven metres without the dubious
benefits obtained from the use of a bicycle). *Figure 37* shows Lavalade in
action, having just launched himself from the ramp. The take-off was recorded
at Juvisy.

The first major gathering of hopeful contestants was held near Paris on
June 2nd, 1912. Twenty-three participants attempted to win the 10 000 francs,
but not one succeeded in even leaving the ground. As a result of this
disappointing effort, Peugeot were induced to donate a second sum for a simpler
task, a flight of one metre at 10 cm altitude. Another racing cyclist, Gabriel
Poulain, succeeded in winning this prize on 4th July, 1912. His distances in
each direction were 3.6 and 3.33 metres, and this is claimed to be the first
true success of a 'flying bicycle'. Poulain was to become one of the most
popular competitors in the many further attempts at improving upon this
performance.

The year 1912 saw the inducement of further prizes and attempts at winning
the major Peugeot prize (*Fig. 38*). The Paris newspaper *La Justice* offered
100 000 francs to the first person to fly from Paris to Versailles and back
non-stop using muscle power. Their money was very safe in their own pockets.
A less ambitious task was set by Dubois, who donated a sum of 500 francs for
what was to be known as the 'Decimeter Prize'. This, for a very short flight,
was won by the German racing cyclist Rettig on the 19th October, 1912, at the
Prince Parc Course near Paris. Michelin, the tyre manufacturers, put up 2000
francs for the completion of a flight of five metres. This was won by Didier,
who was to become another star of flying bicycle events, on the 21st December,
1912. His best run was 5.32 metres. Meanwhile another spectacle involving
competition for the Peugeot 10 000 franc prize had been organised, and took
place on 24th November, 1912. Again efforts to leave the ground were abortive,
seventeen participants making attempts on this day.

In the same year further attempts were made by a number of the more
enthusiastic competitors, notably Poulain, Rettig, Didier and Lavalade, to win
the Peugeot prize. A variety of small prizes were won for minor successes,
but the ten metre flight seemed no nearer accomplishment than during the
previous year.

The machines entered for the Peugeot competition could be divided into two

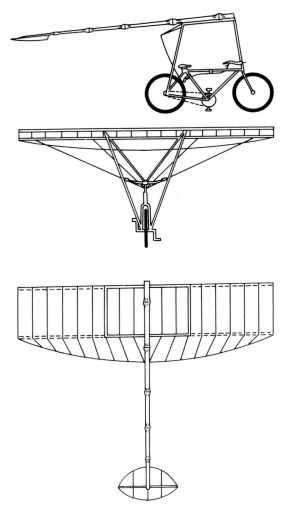

*Fig. 38* The Ladougne 'Aviette' of 1912.  The wing span
was 5 metres.

categories, those which relied upon pure momentum for their flight, and the
more sophisticated flying bicycle which sported a propeller or flapping wings
in an attempt to sustain flight once it had become airborne.  Using the basic
bicycle with fixed wings, the approach to a flight was to generate as much
speed as possible and jump off a slight incline, as done by Lavalade in May of
1912, or to use wing lift to leave the ground.  The practice was generally to
increase the wing incidence just prior to take-off, hoping that a sufficient
forward speed had been reached.  These relied, of course, solely on a glide
once take-off had been achieved, and only with sustained effort could the term

'man-powered aircraft' be truly appropriate.  The French word 'Aviette' was
coined to describe these aircraft/bicycle combinations.

As is to be expected in a competition of this type, there was a large
proportion of entrants who could be described, to put it kindly, as eccentric.
However, some undoubtedly serious attempts were made to design what were
thought at that time to be practical man-powered 'Aviettes'.

Some of the more simple machines were tested without wings, aided solely
by pusher or tractor propellers to increase ground speed.  This proved an
ideal way for measuring propeller efficiency, but apart from high speed jumps
off inclines, no take-off could be expected.  Most of the winged machines
tended to be monoplanes, primarily because it was thought that the biplane would
be too heavy.  The monoplanes of Manuel Gracia, Moulin and Groussy were designed
to successfully accomplish very limited jumps, and had no airscrew.  Wings were
mounted directly below the handlebars of conventional bicycles, the average
wing span being of the order of 4 metres.  Some designers incorporated a
rudder/fin assembly to assist stability, but movable control surfaces as
such were rarely used - generally the 'flight time' precluded any calculated
adjustments.

Larribe attempted to win the Peugeot prize using a tandem high wing layout,
the two wings being of identical size mounted above the head of the pilot on
pylons.  This layout was also used by Vincent de Montrouge, whose aircraft was
reported to partially leave the track during an attempted take-off on level
ground.  In *Fig. 39* however, the rear wheel can be seen to be neatly crushing
the cross wires over which the 'Aviettes' were supposed to fly.  Although
neither of these machines had rudders for direction control, de Montrouge's
aircraft shown in *Fig. 40*, was aided by a pusher propeller driven by a chain
via the wheel hub.

Hassay, an employee at the Voisin aircraft factory in France, constructed
a canard biplane.  The upper mainplane had representative ailerons and the front
bicycle wheel was replaced by an extended frame carrying the front wing.  Two
dolly wheels were used to ensure stability on the ground, and supported the
light fuselage.  No airscrew was used, and provision for lateral control was
omitted.  In order to prevent damage to the lower lifting surfaces and the rear
fuselage, two tail skids were attached to the airframe.  This aircraft did not
succeed in leaving the ground.

The successes of Didier have been referred to previously in this chapter,
and a description of the machine used in his 1912 flight is appropriate.  It
originally consisted of a conventional bicycle to which was attached a pair of
elliptically shaped flapping wings, each wing being about 1.5 metres long.
Before an attempted flight the wings were set with an extremely large dihedral
angle of about eighty degrees *(Fig. 41)*.  When Didier judged that he had
sufficient speed to take off, the wings were released and brought down to a
horizontal position using elastic rubber cables.  No flights were made with
this method.  Influenced by the simple design of Rettig, who won one of the
'consolation' prizes on the 10th October, 1912, using two short aerofoils as
auxiliary lifting surfaces, Didier modified his machine.  He removed the
flapping wings and replaced them with devices similar to those employed by
Rettig, while retaining some ability to move these shorter aerofoils.  Two
auxiliary lifting devices were also located close to the rear wheel.  With this
machine Didier won the Michelin prize for a five metre flight on December 21st,
1912, at the Prince Parc, Paris.

Yet a further meeting was held in June 1913 in Paris, when approximately fifty competitors met to attempt the Peugeot course. This proved to be one of the most disappointing meetings, none of the participating aircraft managing to leave the ground. Most of the machines were propeller driven monoplanes, and of these, Perray's was one of the more complex designs.

The Perray machine was a small monoplane, with a bicycle providing the basic ground propulsion unit. However, in transferring the pedal drive to the propeller three separate chain systems were used, the propeller being in a tractor configuration mounted in front of and above the handlebars. While on the ground the rear wheel was driven in the normal manner, and at the same time the propeller was operated via a chain leading from the wheel hub to a large sprocket wheel slung beneath the cross bar. A chain led from this wheel to a bevel gear adjacent to the propeller shaft. The front steering wheel was provided with a fairing which may have been to reduce windage losses, or to protect the wheel from damage should the propeller strike it.

One of the few biplanes present, that built by Givaudan, also had an unusual transmission system. The pusher propeller was operated by a treading motion, similar to that on a contemporary four wheeler toy, where the two front wheels were steered by feet and the forward drive generated via a form of handlebar crank pushed forwards and backwards, connected to the rear wheels. The structure used by Givaudan was relatively light and seemingly insecure, but the opportunity to test this in flight never materialised, the machine only making a rather erratic run down the race-course. Didier also attended, and made a further attempt, which was unsuccessful in spite of the assistance of a pusher propeller of comparatively enormous proportions.

There were several ornithopters at this meeting, one was a bicycle with flapping wings to which the pedals were connected by lever arms. La Wera also used an ornithopter, and achieved a ground speed of 15 km/h. Other machines are shown in *Figs. 42, 43 and 44.*

In order to encourage the competitors, and the spectators, who were no doubt discouraged by the lack of flying success, the promoters offered three prizes of 2000, 1000 and 500 francs for the fastest times to reach a speed of 40 km/h. Seventeen competitors in all took part. René Bernhard reached the required speed in 30.2 seconds. Charles Dieudonné and Mouille received the second and third prizes with times of 31.2 and 33 seconds respectively.

This event was the final 'Peugeot Prix' meeting before the First World War, and although success had eluded all competitors, much had been learned. Sceptics however, were still predominant, believing that short jumps were the greatest flights likely to be achieved using muscle power. It took another eight years to resolve the argument.

In June 1920 the rules for the Peugeot prize were revised, although the basic aims remained effectively the same. The most interesting additions to the rules were as follows:

(i)  The Peugeot prize of 10 000 francs in cash was to be awarded to the first flyer who successfully passed this double test[1]

---

[1] *The 'double test' refers to the need to complete a flight in both directions (Section 3 of the original rules).*

during the 'Aero-cycle' weeks which were to be held in Paris
from 23rd - 31st April, 23rd - 30th June, 24th - 31st July
and 23rd - 30th September, 1921.

(ii)  Each competitor was to be permitted a period of fifteen
      minutes from the time of start to carry out his flight in
      both directions.  The start was defined as the moment the
      machine completely crossed the first line, whether it be in
      flight or taxiing.

(iii) All those competitors who had successfully completed a flight
      thereby directly qualified for a new prize, that for the
      greatest distance flown, to be awarded immediately after the
      main event.

(iv)  In the eight days following official recognition of the
      winner of the Peugeot prize, the organising committee were
      to publish a new set of rules, in which was to be stated the
      time and conditions according to which this second prize was
      to have been awarded.

To supplement the prize-money a donation had been made by George Dubois,
to be awarded to the first pilot to reach an altitude of ten metres with his
machine.  This prize, originally 500 francs, was increased to 700 francs by
further small gifts.  It was stipulated in the rules governing its award that
the flight at ten metres altitude should be horizontal for a distance which
remained unspecified.  The competitor was also free to use sloping ground, or
any other assistance he desired, to aid take-off, provided that a motor was not
used and no part was added to, or jettisoned from, the machine during the take-
off and subsequent flight.

The journalists of the day favoured the machine entered by the Mauve company,
and the one constructed by Louis de Monge, a professional engineer.  Monge's
aircraft, flown by Abbins, had, it is claimed, reached a height of eight metres
during a flight of the same length at 40 km/h in 1919.  This machine had an
empty weight of 34 kg and a wing area of 13 $m^2$.  A number of well-known powered
aircraft manufacturers of the era were represented in the competition, Nieuport
and Farman being the most noteworthy.

## GABRIEL POULAIN

Much was written in the press regarding the success of Gabriel Poulain who,
on 9th July 1921, won the Peugeot prize of 10 000 francs.  The predictions
made by the French journal *L'Auto* on 9th January 1912 are worth recording at
this stage:

"Although it has not yet been demonstrated experimentally, we remain
convinced that the aerocycle, even without a propeller ... will completely
leave the ground and without great difficulty be capable of making the
minimum jump of ten metres, which will beat the length jump record.  If a
propeller is added the machine will be improved, and if the skill and
experience of the rider increases, then one record after another will be
broken.  In this way ultra-light aviation will develop alongside large
scale aircraft, although not at the same altitude.  The aerocycle below and
the aeroplane above;  thus both will fly without interfering with one

another.  There is plenty of space in the sky and there is room for both.
Room even for three, because between these two a third will be introduced,
just as the motor-cycle has taken up a position between the bicycle and
the car.  But before we talk of a motor powered aerocycle, we must first
construct the  man-powered aerocycle."

*L'Auto* took pride in recording the success of Poulain, and on 10th July,
1921 Emanuel Aimé wrote:

"Others as well as myself will remember the glory of the historic morning
of 9th July, 1921, which brought us the test flight of the aero-cycle,
piloted by Gabriel Poulain.  All the conditions were favourable to guarantee
the triumph of the first aeronaut:  the racing cyclist in good form and in
full control of his powers;  the perfection of the machine constructed by
the Nieuport Company under the direction of Chasserio, Maria, Bazaine, and
Gradis;  favourable weather conditions;  a sympathetic attitude on the part
of the representatives of the Paris press;  and an enthusiastic though
nevertheless cautious crowd of on-lookers, in the first row of which was
the donator of the prize of 10 000 francs, Robert Peugeot, accompanied by
Jean Koechlin."

Poulain made a number of attempts at the Peugeot course on that day.  The
crowd of approximately one hundred had assembled by 3.45 a.m. at Longchamps,
and by 4.43 Poulain was prepared for his first take-off run.  Following the
release of his wings, he succeeded in flying 11.98 metres at a height averaging
1.5 metres, and making a safe landing.  The rules stipulated that a flight must
immediately be attempted in the opposite direction, and Poulain achieved 11.59
metres at a slightly lower altitude in this stage.  However, a misunderstanding
about the starting conditions, whence take-off must be between two lines,
nullified this attempt, and preparations were begun for a repeat run.

The second series of tests proved satisfactory, flights of 10.54 and 11.46
metres at an altitude of over one metre being made.  Take-off on the first run
was timed at 5.54 a.m., and the return run commenced at 6.03 a.m.  This
officially observed and recorded set of flights resulted in Poulain being
awarded the Peugeot prize.

Poulain's machine had a wing area of 12.08 $m^2$, being a biplane with an
upper wing of span 6 metres, chord of 1.2 metres, the lower wing having a span
of 4 metres with a chord of 1.22 metres, *(Fig. 45)*.  Stagger amounting to 0.7
metres was employed, and the distance between the wings was 1.2 metres.  At the
time of the successful attempt on the Peugeot prize, Poulain weighed 74 kg and
his aircraft 17 kg.  As well as being a racing cyclist, it is interesting to
note that Poulain was a qualified powered aircraft pilot, having done a
considerable amount of flying before the First World War.  Details of the flight
were recorded in a report signed by the aerocycle experts and two copies were
written on parchment and donated to the 'Conservatoire des Arts et Métiers' and
the Carnavalet Museum.

Poulain himself wrote profusely on his record attempt and the development
of aviation as he saw it, and, no doubt influenced by his successful flights,
made various forecasts which with hind-sight, one is tempted to ridicule.  His
writings did contain some indication of a simple but very logical thought
process.  One of his pieces, titled *The Development and Future of Aviation*, is
typical.  A rather literal translation reads thus:

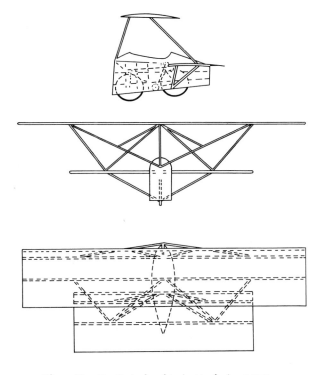

*Fig. 45* Poulain's 'Aviette' in 1921

"When I (Poulain) made my first attempt with the aerocycle most people
regarded me as a fool or at least an eccentric.  I had faith, because I
had already devoted ten years of study and because all my hopes had been
realised.  But success does not come overnight!

What I wanted above all else was to take-off, because without a take-off
there can be no flight.  I am sure all the academics will agree with that!
But they were not directly involved with this (Poulain's) work.  All of
them derived Proofs, with A plus B, that can never, absolutely never, lift
a man into the air by his own power ... I believe that if I had understood
anything at all of their calculations I would have been tempted to drop my
research:  they would have got the better of me.

I firmly believe that whatever may be said to the contrary, the aerocycle
is destined for a great future.  It will certainly not be used like an
aeroplane, which would make it superfluous.  I can however see it as a
cheap little machine which will enable us to travel on the road;  suddenly
to avoid a long detour one will fly over a depression and a few kilometers
further on land again on the right track.  When will it be developed to
this stage?  I don't know, but were people able to look forward to the
future, and predict what would subsequently happen to the aeroplane?

People claim that I can fly because I am a cycling champion, and others

would not be able to achieve my success.  In the first place there is
nothing to prove this.  One must then ask;  would someone else have flown
the Channel with Bleriot's machine, or have lifted themselves into the air
as the Wright Brothers did?  If anyone could do it, then where does the
merit lie?

It is also said that I will never do any better with my aerocycle, which is
capable of flying ten metres.  I fully agree with them.  With this machine
I only wished to demonstrate that one can take-off;  I had no further
ambition.  But this proved that lift was present, which showed that it
would later be possible for me to maintain direction, gain height, and
catch the up-current necessary for keeping me in the air.

My next aerocycle will obviously be fitted with a propeller."

In 1922 Poulain commented on the proposed competitions for the new Peugeot
prize of 20 000 francs.

"When in July last year I had the satisfaction of winning the Peugeot prize
with a flight in two directions over a distance of ten metres, as required,
... the donator of the prize, M. Robert Peugeot, offered a sum of 20 000
francs for another attempt.

The sporting experts gathered to work out a set of rules, which at least
places on record their unqualified faith in the possibilities of man-powered
flight ... They argued:  if a competitor has to fly ten metres for 10 000
francs, then for 20 000 francs he would have to fly twenty metres.  This
was logical, but it is more important to understand that one could ask for
more (a greater distance);  for the ten metre flight the most difficult
stage was to demonstrate the possibility of a take-off.  Once I had shown
this was possible, the increase in the duration of flight could be the next
goal."

Poulain went on to voice his objections to the rules for the 20 000 franc
Peugeot prize, for which the competitors had to make a continuous flight of
fifty metres over a flat course, shortly to be followed by a flight over the
same distance in the opposite direction.  Pilots were allowed twenty minutes
to complete both flights, and three attempts were permitted to each participant.
Poulain commented thus:

"If it is only a case of an aerobatic jump, why then demand two flights of
fifty metres in immediate succession in less than twenty minutes?  Do they
want to destroy the spirit of the inventors and also their enterprise?
The encouragement of progress is a strange art.  We must recognise, we poor
bird-men, that we have less protection than the pilots of the motorless
aeroplane (glider?), the future of which is far more hypothetical than that
of our aerocycle.

They (the governing committee) will concede that if an aerocycle flyer
covers fifty metres, then his performance is worthy of interest.  Even if
it is assumed that he were assisted by a cyclonic wind, if he did succeed
in making this 'jump' the power required would be somewhat different to
the initial power during take-off.  A great success would have been acheived.
Why then further increase the difficulty and stipulate a return flight?
And the most difficult thing is to make that return flight less than twenty
minutes later!  When taking off we have to make a considerable effort.

I do not want these remarks to be seen as the accusations of a bird-man seeking unjustified excuses or evasions.  I still continue to have great faith in my next experiments.

My new machine, constructed by the Nieuport Company under the management of Marie and Chasserio, the men who, with the assistance of Delage, made possible my success in 1921, inspires us with great confidence.  I will content myself with a few brief details of it:  a propeller will enable me to stay airborne;  the machine will be a monoplane with a wingspan of six metres.  The all-up-weight will be 100 kg and I expect to develop two horsepower (1.5 kW) for take-off[1].  Experiments with a propeller alone have convinced me that I can attain 40 km/h despite the drag of the wings attached to my bicycle."

Sadly, Poulain was not to achieve success with this machine, in fact it is doubtful if any trial runs took place, and the 20 000 franc Peugeot prize lapsed into history.

* * *

The 1928 edition of *Who's Who in Aviation* includes the following entry:

"BARTLETT, Reverend Alfred James

b. Watton-at-stone, Herts, Aug. 15, 1866;  s. of the late Rev. J.B.B. (Rector of Rowberrow, Som.).  Club:  Westminster.  Made monoplane glider model embodying the dipping entering edge, 1887;  attempted design of light rotary steam engine for direct lift machine, 1892;  testing experimentally ... types of foot propellers for gliders.  Add.  The Rectory, Marston Magna, Nr. Yeovil, Som."

The Rev. Bartlett was noted for his 'high wire act' in which he furiously pedalled a propeller-driven machine, generally wingless, suspended on a wire. Photos taken from a convenient distance suggested no visible means of support: This perhaps is symbolic of many of the less sophisticated attempts at man-powered flight in this period, soon to be eclipsed by work in Europe.

* * *

---

[1] *It is quite possible that Poulain could have generated 1.5 kW at the moment of take-off, but it is highly unlikely that he would have been able to maintain such an output for more than a few seconds.*

*Fig. 25.* Otto Lilienthal in flight, 1896.

*Fig. 26.* Frost's second ornithopter, although not
man-powered, is of interest because of its
sophisticated wing structure.

*Fig. 27.* The Keith-Weiss machine shortly after take-
off.

*Fig. 28.* The Keith-Weiss 'Aviette' in August 1912.

*Fig. 29.* Towed launch of Palmer's fourth prototype at
Cleeve Hill, 1921.

*Fig. 30.* An obscure English attempt at flight in 1920.

*Fig. 33.* Activity in the United States in the 1920's was typified by the machine constructed by Lehman Weil, photographed in 1927.

*Fig. 34.* W.F. Gerhardt's multiplane photographed in July, 1923.

*Fig. 35.* An 'aircraft' constructed by A. Schmutz in
1904, and exhibited in France.

*Fig. 36.* The Piat machine, one of the more sophisticated products of 1912.

*Fig. 37.* May 1912 – Lavalade and ramp-assisted take-
off.

*Fig. 39.* The Vincent aircraft during an abortive
attempt on the Peugeot prize.

*Fig. 40.* M. Vincent and his 'Aviette' at the moment of
its collapse following his jump, in 1912.

*Fig. 41.* Didier and his 'Aviette', late 1912.

*Fig. 42.* A propeller-driven monoplane of unknown
origin, dated 1913.

*Fig. 43.* Another 'Aviette', the wings of which could
be folded for transport.

*Fig. 44.* On 2nd June 1912, Mons Malby failed to take
off on this tricycle.

# 5

# Pre-War Germany

*– the 'Muskelflug Institut'*
*and Haessler's 'Mufli' aircraft*

*5*

## GERMANY 1920 - 1938

Interest aroused by the Peugeot 'Aviette' Competition was no doubt partly responsible for the initiation of the Rhön contest in Germany in the mid 1920's. Although this event attracted much less popular interest than the mass attempts at the Prince Parc, Paris, the entries included several machines which we would possibly recognise today as comparatively sophisticated designs. (The next decade was to herald an even more rational and calculated approach to man-powered flight, particularly within Germany, and current thought has been greatly influenced by the work of Ursinus, Haessler and others).

One machine intended for entry in this competition, but which did not fly at the time, was that constructed by Porella in Beuthen, Upper Silesia. The aircraft had a unique system of rotating aerofoils, and sufficient interest was aroused by Porella's proposals for the designer to be given RM 300 by the organising committee of the Rhön competition to assist him with the construction of his aircraft. The aerofoil system resembled the paddle-wheels of ships, but as well as providing thrust it was designed to make a major contribution to lift. As the rotating aerofoils, which were mounted on arms of approximately one metre radius, approached the downstroke they assumed a horizontal position, thus generating lift from both the forward motion of the aircraft and the downward movement of the aerofoil. Once the half cycle had been completed and the aerofoil was at its lowest position, the chord line rotated through 90 degrees so that the line of action of the lift was in the direction of motion of the aircraft, providing thrust for forward flight, *(Figs. 46 and 47)*.

A fixed wing was provided on top of a small pylon in front of the rotating wings. The wing span was 10 metres with an area of 9.5 $m^2$ while the rotating wings had an area of 6 $m^2$. Conventional tailplane and rudder surfaces were used, with areas of 2.5 and 1 $m^2$ respectively.

No reports are available on the performance of Porella's aircraft, and it is not known whether construction was ever completed. The available illustrations suggest that the pilot would have had considerable difficulty in pedalling and controlling the aircraft in the position shown.

Another attempt to use a similar form of propulsion to that proposed by Porella was made by Josef Helbok, an Austrian mill-worker from Höchst. Helbok was an enthusiastic model-maker, and obtained most of his knowledge about the rudiments of flight from carrying out small experiments over a period of fourteen years. His machine, constructed in 1932, was driven with the aid of a rotating surface mounted ahead of the wing. This surface was S-shaped, and was designed to provide forward thrust. The accelerating air produced by this aerofoil was then passed over the fixed wings to aid lift. Helbok's aircraft was a tail-less biplane in configuration, with a total wing area of 36 $m^2$.

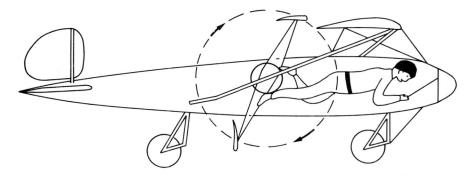

*Fig. 46.* The aircraft proposed by Porella in 1925.

The empty weight of 65 kg was rather high for a machine of this size, but one advantage Helbok had over Porella was in the position of the pilot, the upright position adopted by the former being more comfortable and probably much more efficient. Stiasny[1] suggests that a flight of fifty metres was achieved (at an average altitude of two metres), with Helbok piloting. A pair of motorcycles towed the aircraft to gain acceleration during the take-off run. Unfortunately the machine was destroyed during a later towed launch, and lack of financial backing caused the project to be abandoned.

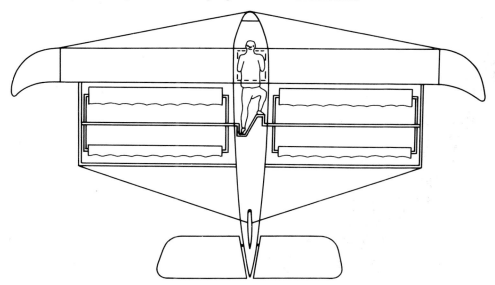

*Fig. 47.* A plan view of Porella's 'paddle-plane'.

---

[1] *H-G. Schulze and W. Stiasny. 'Flug durch Muskelkraft'. Verlag Fritz Knapp. Frankfurt A.M. 1936.*

## JAKOB GOEDECKER

As well as the rotational and flapping motion of wings in isolation, some designers attempted to utilise a combination of both, using arguments based on crude observation of bird flight. Working along these lines, the aircraft constructor Jakob Goedecker, apparently well known in the pre-war years, built in 1929 a free-flight model, 'Bird I', with which he achieved some success. This machine, which had wings in the form of a fairly steep V, weighed 414 g and had a 0.5 W rubber motor which provided the power necessary for horizontal flight. Whereas 'Bird I', surprisingly, not only flew in a stable manner, but also changed over to equally stable gliding flight when the wings were stationary, Goedecker's other models, which were constructed on the flapping-wing principle, were less satisfactory.

Writing in a 1934 issue of *Flugsport*, Goedecker said:
"My model 'Bird I' was driven by stretched rubber and rotating wings which moved in a conical path. With this form of drive neither the weight of the wings nor their wing-span caused any special difficulties; on the other hand the wings operate with a much lower driving force and their movement has no disturbing effect on the stability of flight. But since nature does not employ this rotary wing motion in a long flight, I have tried in the case of other models to endow the wings with an upward and downward flapping motion. Although the wings and crankshaft of my 'Sea Swallow' are very light indeed, I have not succeeded in making more than three wing-beats in flight. Usually the model did not execute more than one beat. Some special factor must have been involved, as a result of which, when the rubber motor was fully wound up, the wings could only overcome the top and bottom reversal points twice. I then attempted to correct this fault by attaching a flywheel, and for this purpose constructed the model 'Seagull'. So that the flywheel would have the maximum kinetic energy for a low weight, it was driven directly by the high-speed rubber motor, while the wings were driven by a slow-speed crankshaft. All parts were mounted in ball-bearings and friction was very low. The result was that the wings were only able to make ten flapping motions, and I noticed that at the reversal points the wings did not flap like the wings of a bird, but performed a very strange slow reversal of the direction of motion. In both the models referred to above gliding flight is stable, even in flapping flight, but in the 'Seagull' model no forward propulsion was noted. It was clear to me that there were faults in the mechanism for actuating the wings."

Goedecker then attempted to eliminate this defect by constructing the 'free-coupling' in which the wings and the drive system were not linked by a rigid connection, but in which the free crankshaft was engaged in a sliding link in the wing spar, producing the flapping motion.

Of interest and significance is Goedecker's observation that in the case of the rotating wing principle the drive presented no special problems and the wings worked with a much lower driving force.

Piskorsch also employed this principle as the basis for the design of his rotating wing models. Unlike Goedecker, who used only one set of surfaces, Piskorsch's aircraft had three. These three pairs of cantilever blades mounted in the fuselage on a conical drive were very narrow, and acted like springs each accelerating the air with a beating motion. Piskorsch proposed to construc a man-carrying aircraft in accordance with these principles, but his plans never reached the hardware stage. However, a model was made by the Warzog Coachworks in Troppau, Czechoslovakia, *(Fig. 48)*.

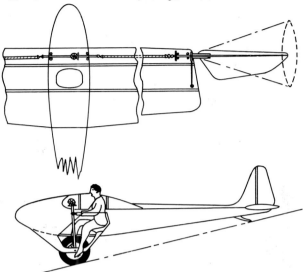

*Fig. 48*  The aircraft designed by Piskorsh in 1934. Note the rotating wing tips.

## LIPPISCH AND BRUSTMANN

Alexander Lippisch was one of the earliest advocates of serious research into man-powered flight. (This assumes that one accepts the German work on man-powered flight in the late 1920s and subsequently as the first logically developed and, in part, successful programme on the topic.) Lippisch claims to have been responsible, via the medium of a lecture he gave at the Lilienthal Gesellschaft, for the technical bias in the Rhön contests of 1926.

An association between Lippisch and Dr. Martin Brustmann which was to continue for many years appears to have formed at this time. Brustmann in the 1920s carried out studies into the power outputs of humans, based on his experience as an expert on medical matters, specifically physiology, related to sport. He produced figures on the short duration power outputs of athletes, stating that a man could produce up to 1.5 kW in a very short burst, reducing to a continuous output of 0.19 kW.

Brustmann proposed means for utilising as many as possible of the muscles in the body, the total weight of muscle being used amounting to approximately 25 per cent of the total body weight. Brustmann suggested using the arm and chest muscles, in conjunction with the main leg muscles, in a combination of arm movement and the oarsman's roller sliding seat. (It is of interest to note here that Everling, writing in 1934, describes how he applied for a patent, in conjunction with Erich Offermann, in 1924. This patent was not granted, but the invention he described was an extension of the ideas put forward by Brustmann. The main feature of the design was a means for over-coming the severe shift in the position of the centre of gravity which is associated with a sliding seat. Such movements would of course be unacceptable in an aircraft, as the pilot would be continuously compensating, using the controls, for the resulting pitch angle changes.

Various claims were made for the total power generating capability of a man using such a combination of muscles. Everling believed that 1.9 kW would be achieved momentarily by a trained oarsman, but Brustmann is reported to have quoted a figure 5.3 kW being attainable for periods of less than ten seconds. This remains the highest value claimed for human power output in the context of man-powered flight.)

A few years later Brustmann, aided by Lippisch, constructed an ornithopter with a wing span of about 10 metres and an empty weight of 30 kg. A young athlete, who was also a pilot, Hans Werner, carried out the first flight trials, which were performed on a downward incline *(Fig. 49)*. After several attempts, a flight of about three-hundred metres[1] was reported to have been achieved. In this distance, eleven flappings of the wing were carried out. It is doubtful whether such exertions extended what was in fact a glide.

Lippisch, in criticising the machine, complains of the inadequacy of the transmission (this being via cables running over pulleys), but heralds the fact that it flew. Following these early attempts, Brustmann took his aircraft to Berlin to give demonstrations. However, he attached a large undercarriage to the fuselage, with the result that the machine was unable to take off! Lippisch then appears to have attempted to further his work alone. In the summer of 1933 he carried out experiments on wings for man-powered aircraft, but his tests were cut short by his enforced transfer to another research establishment at Darmstadt.

The record for truly persistent participation in man-powered flight probably rests with Lippisch; he was still actively involved in the design of such machines in the 1960s.

## OSKAR URSINUS AND THE 'MUSKELFLUG-INSTITUT'

The results of the work of Oskar Ursinus carried out at his Muskelflug-Institut have remained among the most comprehensive and reliable data on the power outputs of human beings. Details of the vast majority of his experiments related to man-powered flight were published in Germany in 1936, and covered measurements made on men in a variety of attitudes from the conventional cycling position to the fully reclining and prone positions. (References to this work are listed in the Bibliography in Appendix IV).

The Muskelflug-Institut (Institute of Man-Powered Flight) was established at Frankfurt towards the end of 1935 by the Polytechnischen Gesellschaft with the full blessing of the German Reichsluftfahrtministerium (Aviation Ministry). The aim was to give advice and assistance to workers developing and constructing

---

[1] *Lippisch refers to the length of the flight of Martin Brustmann as being "about 300 metres". The review of rotary wing aircraft, which includes ornithopters, published by the United States Department of Commerce (Office of Technical Services) states however that "a flight of about 60 feet (18.3 m) was made". The machine is additionally described as "a controversial glider".*

man-powered aircraft, and also to carry out an internal programme of basic research.  The service offered by the Institute was such that an individual designer, who was unlikely to be able to afford expensive test facilities, was free to obtain the information he required from experiments carried out at Frankfurt.  The Director of the Muskelflug-Institut was Ursinus, and the governing body included Dr. Krebs, Professor Wachsmuth, Herr Burggraf, the accountant, and Dr. Gramlich, Secretary.

A number of methods for propelling such a vehicle were tried, including rotating, (as in cycling), reciprocating and rudder movements.  Other variables considered in Ursinus's tests included crank length, speed of rotation or reciprocation, and the influence of the phase shift between arms and legs when both sets of limbs are used together to produce power.  His principal test subject in most of the work was Herr Gropp, an athlete aged twenty-seven who weighed 71 kg and was 1.78 metres tall.  However, a number of tests were carried out using various sportsmen, ranging from a mountaineer to a champion swimmer.

The results Ursinus obtained comparing the power outputs using arms, legs, and a combination of both are shown in *Fig. 50*.  A combination of arm and leg motion gave, as expected, the greatest power output, with a slightly higher

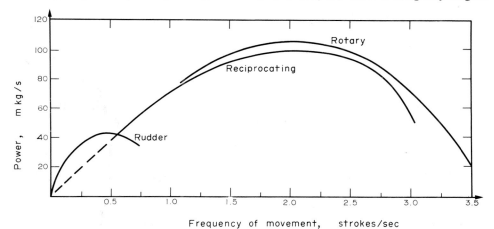

*Fig. 50* The power output of H. Gropp, measured by Ursinus, for rotary, reciprocating and rudder movements.

optimum stroke rate than that required when operating legs or arms alone. The optimum rates, of the order of two cycles per second, are similar to the rates at which racing cyclists produce their best performances.  The rotary motion gives a slightly higher power output capability than a reciprocating motion, but in the applications for man-powered flight it has generally been suggested that a combination of both forms of power generation, in the form of cycling and hand cranking, would be a suitable compromise.  Ursinus found that the optimum stroke length was slightly less than 400 mm for rotary movement.  With reciprocating motion, the phase shift of arms and legs had a noticeable effect on the power output, optimum conditions being achieved when the motions were in phase with one another.

The best results achieved by Ursinus were obtained with a racing cyclist in training. The test subject, P. Hoffmann, was twenty-six years of age and weighed 70 kg. His height was 1.7 metres. Hoffmann achieved a 30 per cent increase in power output over that produced by Gropp; in fact his power output using legs alone was only slightly less than that produced by Gropp when utilising both arm and leg movements[1]. Hoffmann's maximum power outputs are shown in *Fig. 51.*

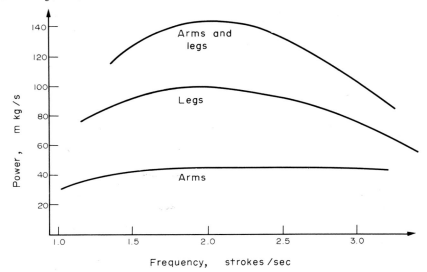

*Fig. 51* The maximum power output of Hoffman, using arms, legs, and a combination of the two.

Familiarity with the motion used for producing power was found to be an important factor when testing cyclists on reciprocating machines and other athletes on cycle simulators. For example when the performance of P. Hütter,

---

[1] *The amount of power produced by the muscles depends on the capacity of the heart and lungs, and the efficiency with which the oxygen, which acts as fuel, can be adsorbed in the lungs and thence taken to muscles through the bloodstream. The heart and lung capacity therefore limits the power output of an individual when measurements are made over extended periods, but muscles store energy, which may be released to augment a burst.*

*It may not always be the case that arms and legs used in conjunction will prove more powerful than legs alone. In tests carried out by members of the Royal Victoria Infirmary, Newcastle upon Tyne, racing cyclists using only legs to generate power were utilising almost all of their lung capacity, therefore additional hand cranking would not have increased their output significantly, unless of course some of the stored energy in the muscles had been used for a short period burst.*

a mountaineer and skier, was measured on the rotating motion machine, it was
found that there was a sharp decrease in performance at the higher cycling
rates, particularly when the feet were used for generating motion. A similar
trend was noted when a racing cyclist was tested on the reciprocating machine
but the comparison made by Ursinus was not truly accurate in that the cyclist
used for measurements on the reciprocating rig was much lighter than Hoffmann.
The power outputs of a wrestler were significant in that the performance of
legs and arms was almost identical at the optimum reciprocating speed of 1.75
cycles per second. With a rudder movement, the most inefficient of the three
motions investigated, Ursinus found that at the optimum stroke rate of 0.45
per second, up to 40 per cent of the total time was required for the reverse
stroke, which did not produce usable power.

Investigations were also carried out on the effect of duration of power
production on the output and optimum stroke rate. For cycling with legs only
it was found the optimum rate of pedalling decreased with increasing time,
as shown in *Fig. 52*.

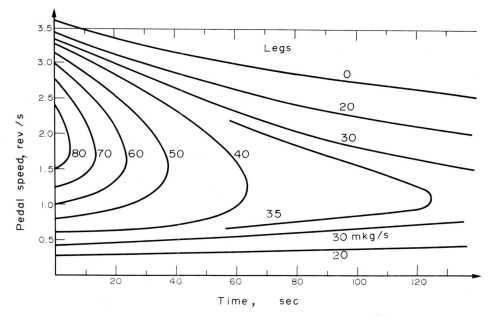

*Fig. 52*  The dependence of human power output on the
stroke rate and duration – pedalling motion.

A crank drive system was shown to be most effective when the pilot was in
the reclining, as opposed to the sitting position. The optimum frequency was
1.7 revolutions per second, and the maximum power output decreased by 50 per
cent after one minute of concentrated effort.

So broad were the terms of reference of the Institute, that, but for the
intervention of World War II, the work which undoubtedly would have been
carried out there could have advanced the knowledge of man-powered flight by

at least twenty years, certainly from the point of view of aerodynamics.
Apart from the assistance given to amateur constructors, research programmes
were initiated to cover the testing of efficient drive units, launching
devices and power storage systems;  research was also planned in subsidiary
fields such as flapping wings, rotating wings with various aspect ratios, and
investigations into bird and insect flight, *(Fig. 53)*.

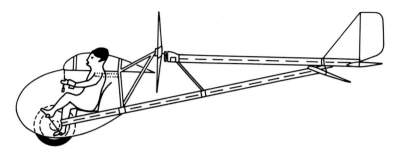

*Fig. 53* An idea for a man-powered aircraft put forward
by Ursinus in 1933.

The work of the Institute was regularly published in the magazine
*Flugsport*, and full details of the only major programme undertaken, Ursinus's
work on power outputs, were spread over several issues.

## THE HAESSLER-VILLINGER AIRCRAFT

Helmut Haessler, a glider pilot, began a series of experiments in 1933
which were to lead (two years later), to the construction of a successful
man-powered aircraft.

His first tests were aimed at determining the power output of cyclists,
(at this time the 'Muskelflug - Institut' had not been established) but the
results he obtained were not particularly encouraging, the test runs using
the fellow members of his gliding club.  His next series of measurements were
made with the assistance of a group of professional racing cyclists, one of
whom offered to assist Haessler with his project.  This cyclist was reported
by Haessler as being capable of developing 1.0 kW for the initial thirty
seconds of any flight plan.

It was Haessler's intention to enter his aircraft in the man-powered flight
competition organised by the Frankfurt Polytechnische Gesellschaft, a prize
of DM 5000 being offered for the first flight over a closed circuit between
pylons 500 m apart.  In this competition, the use of stored energy was
permitted.  The type of energy storage device was to be approved by the
technical committee sponsoring the competition, and had to be carried within
the aircraft.  The other conditions were in many ways similar to those in the
Kremer competition.  The requirements demanded turning around the two pylons,
which in practice meant a flight of at least 1200 metres, including two 180
degree turns, which have proved impossible even with current sophisticated
machines.  Haessler and his colleagues remained of the opinion that such a

test was unrealistic.

At first Haessler was reluctant to comply with the regulation concerning the use of energy storage devices, as this would mean an increase in weight of about 9 kg.  Rather than abandon his project completely, Haessler concentrated on a design which he hoped would be capable of achieving a horizontal flight under man power alone, and he regarded the flight duration and distance to be secondary considerations at that stage.  In order to aid construction of his aircraft, Haessler was joined by Franz Villinger, a colleague at the Junkers factory where he worked.

The Haessler-Villinger 'Mufli' was a single seat monoplane with a layout similar to that of typical gliders of the period, with the addition of a pylon above and forward of the wing leading edge, on which was mounted the pusher propeller, *(Fig. 54)*.  The overall fuselage length was approximately 5.5 metres with a wing span of 13.5 metres.  Wings were braced using wires in an attempt to reduce the weight of the wing main spar.  The pilot was in a reclining position and used his legs, via pedals, to drive the propeller,

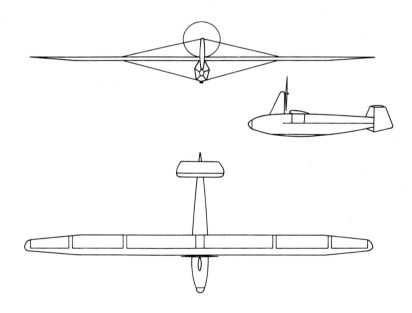

*Fig. 54* Arrangement drawing of the
Haessler-Villinger 'Mufli'.

Further data on the Haessler-Villinger machine is given in Table 1.

The drive from the pedal wheel to the propeller was by means of a belt, and as this belt had to be twisted through 90 degrees it had to be both flexible and have a high friction coefficient, to minimise slip.

In the design of the belt, life was a minor factor.  Fabric belts were

developed with a rubberised coating, and the pulleys also were covered with a
layer of rubber on a metal base.  As a result, Haessler was able to claim a
belt efficiency greater than that of a straight chain drive[1].  Problems caused
by the stretching of these belts, visible in *Fig.56*, necessitated their
replacement following every six flights to ensure maintenance of a high
efficiency.

Table 1.  Haessler-Villinger 'Mufli' aircraft

| | |
|---|---|
| Wing Area | 9.65 m$^2$ |
| Wing Span | 13.5 m |
| Wing Loading | 11.5 kg/m$^2$ |
| Aspect Ratio | 18.8 |
| Wing Section | Gottingen 535 |
| Rib Spacing | 20 cm |
| Wing Chord | 76 cm |
| Sink Speed | 0.52 m/s |
| Propeller Diameter | 1.5 m |
| Propeller rev/min | 600 |
| Propeller efficiency | 82 % |
| Empty weight | 34 kg |
| Energy Accumulator wt | 10 kg |
| Pilot weight | 65 kg |

    Subsequent criticism of Haessler's use of belts, based on the argument
that they are inherently inefficient, was refuted by him.  He compares the
overall propulsive efficiency of the Bossi-Bonomi aircraft, described in
Chapter 6, and Nonweiler's proposed machine, detailed in Chapter 8, with
'Mufli'.  His figures are given in Table 2.

    Haessler obtained the figures in Table 2 on the assumption that to determine
the total efficiency of a propulsive system, one must consider not only the
propeller efficiency and frictional losses in the drive system, but also the
power necessary to keep the propulsion system aloft, and he assumed a value of
8.3 W per kg weight for this power.  Haessler also took into account the extra
structural weight resulting from the need to support the propulsion system.  He
assumed a value equivalent to 80 per cent of the weight of the propulsion
system.  In Table 2 this is called the indirect weight.

    The data presents the Haessler-Villinger aircraft in an extremely favourable
light, but the weight of the drive system, and thus the indirect weight of the
propulsion unit, are underestimated.  For example, the weight of the pylon
supporting the propeller has apparently been neglected in the case of the
Haessler aircraft, whereas a figure representing part of the weight of the
tail fin, which supports the propeller of the proposed Nonweiler aircraft,

[1] *It is interesting to compare the efficiencies of various transmission
systems, using the data recently obtained by S. S. Wilson at Oxford
University.  The results are given in Chapter 15.*

but also contributes to control, has been included[1].

A conventional glider-type control column was used in the rather cramped cockpit of 'Mufli' and normal movements were retained for longitudinal and lateral control. For rudder movement the wheel rotated about a vertical axis, and it was found that the pilots soon became familiar with the control arrangements, *(Fig. 57)*.

Table 2  A comparison of the propulsive efficiency of the Haessler-Villinger aircraft with two other machines.

| | | Haessler-Villinger | | Bossi-Bonomi | | Nonweiler | |
|---|---|---|---|---|---|---|---|
| Weight of Propulsion System | Direct Weight kg | 1.82 | | 16.3 | | 15.9 | |
| | Indirect Wt. kg | 1.46 | | 13.0 | | 21.8[1] | |
| | Total kg | 3.28 | | 29.3 | | 37.7 | |
| | Power required (W) 8.3 W per kg | 27.2 | | 244 | | 313 | |
| | | Effic-iency % | Loss % | Effic-iency % | Loss % | Effic-iency % | Loss % |
| Short Duration Efficiency 0.76 kW | Propeller | 82 | 18 | 87 | 13 | 88 | 12 |
| | Power Drive[2] | 100 | 0 | 94 | 6 | 94 | 6 |
| | Weight | | 3.6 | | 32.2 | | 20.8 |
| | Total | 78.4 | 21.6 | 48.8 | 51.2 | 61.2 | 38.8[3] |
| Long Duration Efficiency 0.38 kW | Propeller | 82 | 18 | 87 | 13 | 88 | 12 |
| | Power Drive[2] | 100 | 0 | 94 | 6 | 94 | 6 |
| | Weight | | 7.2 | | 64.4 | | 41.6[4] |
| | Total | 74.8 | 25.2 | 16.6 | 83.4 | 40.4 | 59.6 |

1. Includes 50% of fin weight (9.1 kg).
2. Efficiency of bicycle chain assumed 100%.
3. Loss related to 1.5 kW (2-seater).
4. Loss related to 0.76 kW (2-seater).

[1] *Lippisch's comments on the Haessler-Villinger aircraft include criticism of the transmission system as being inefficient and the fact that the propeller diameter was too small. He also thought the Gottingen 535 wing section too thick, and in fact Lippisch, in recent correspondence, considers that a 180 metre flight was an amazing achievement with such a machine. Lippisch concedes the point that one could see the aircraft climbing slowly (under man-power). Lippisch takes credit for "persuading Ursinus to work on power outputs" and thence initiating the German man-powered flight contest, and a reference to Lippisch's early work on man-powered flight appears earlier.*

In an attempt to save weight, separate control surfaces were avoided where possible. For lateral control the angle of incidence of the wing semi-spans could be changed differentially, and for longitudinal control the angle of incidence was changed symmetrically. During flight trials it was found that lateral control was sufficient, provided that the aircraft remained in an almost straight and level attitude. The control of incidence angle was poor in that some very steep climbs could result during take-off, causing stalls. Partly because of this, the control surfaces were later replaced with conventional elevators and rudder.

Construction of the Haessler-Villinger machine followed accepted techniques of the period, and the choice of materials was limited. It would appear that the design philosophy was simply equivalent to stripping a glider of similar overall dimensions of all unnecessary instrumentation, restressing the structure for operation in a limited flight envelope with a limited payload, and replacing the cantilever wing with a braced section. Starting with a glider of 150 kg empty weight, Helmut Haessler produced a man-powered aircraft with an empty weight of 34 kg, *(Fig. 58)*. The reduction in weight when compared to that of an equivalent glider is considerable, but this design philosophy neglects the different aerodynamic requirements of a man-powered aircraft, which are detailed later.

The resulting design had a high ultimate load factor of six, mainly brought about by the requirement that wing deflections should be minimised. Cedar plywood was used for many of the major load bearing components, and thicknesses as low as 0.6 mm were found satisfactory for some members. The wing and tail surfaces were covered with silk, and after doping this weighed 160 g/m$^2$. It was found important to apply all lacquers etc. using a spray, minimising the chance of over-application. The fuselage was covered with 0.6 mm cedar plywood, but the gauge was increased around the pilot position. The use of bracing wires considerably assisted in reducing main spar weight. (The energy storage system was incorporated at times, and in order to maintain the stability of the aircraft when auxiliary power was not used, an equivalent weight of 9.7 kg was generally carried in the aircraft nose).

At this stage it is of interest to outline the flight techniques and achievements of the aircraft, enabling these to be viewed in perspective when the performance of subsequent man-powered aircraft is discussed.

As mentioned at the beginning of this section, Haessler had intended to train a professional cyclist to pilot the aircraft, but his plans were not to come to fruition as the cyclist concerned was observed by Haessler to be much more interested in bettering his professional performances than in devoting time to learning to fly!

Haessler carried out some experiments in which he determined the power output of a man, but the more comprehensive work of Ursinus at the Muscle Flight Institute, reported elsewhere in this chapter, and indeed most of Haessler's results, were obtained too late to influence the design of the Haessler-Villinger aircraft. Villinger, writing in 1960, refers to this:

"Unfortunately these purely theoretical data (Lippisch and Brustmann power values) were used in the design of our 'Mufli', the main design basis being a duration of two minutes. It was not realised until our own tests and those of the Muscle Flight Institute, which was founded later, had been done, that the earlier data gave more than double the actual power available.  For this reason the prize was not won by our aircraft ..."

Data obtained by Haessler led him to conclude that by pedalling, a professional cyclist could achieve an output of about 0.35 kW over an extended period.  (The corresponding outputs of a trained amateur and an average cyclist were 0.28 and 0.21 kW repectively).

The first flights under man-power of the Haessler-Villinger machine were made at Rebstock airfield, near Frankfurt a.M. on August 29th, 1935, and a total of seven flights were made on this and the subsequent two days.  These performances are detailed in Table 3.

Table 3.  A record of the early flights of the Haessler-Villinger 'Mufli'.

| Date of Flight | Time | Distance m. | Duration sec. | Height m. | Comments |
|---|---|---|---|---|---|
| 29/8/35 | 11.10 | 120 | 17 | 1 | Slight damage. |
| " | 18.22 | 195 | 20 | 1 | |
| " | 18.43 | 177 | 18.5 | 1 | |
| 30/8/35 | 07.45 | 235 | 24 | 1 | |
| " | 08.14 | 150 | 14 | – | Stall, damage to fuselage nose. |
| 31/8/35 | 17.55 | 220 | 21 | 4-5 | |
| " | 18.44 | 204 | 20 | 4-5 | |

These flights were recorded by three official observers, Oberleutnant Hartog, Hauptmann Jensen and von Lechner, and were generally hailed within Germany as a great achievement.  On August 30th, 1935, Oberst Loerzer the Director of National Sporting Aviation was prompted to write the following note to the Reichsminister der Luftfahrt, Air Marshal Goering:  "Yesterday and today the first man-powered flights took place in Frankfurt a.M.  Pilot Düennebeil[1] flew 235 metres at an altitude of 1 metre.  The machine was constructed by two engineers, Haessler and Villinger." *(Fig. 59)*.

The results obtained during the three days enabled the designers to gauge modifications needed to increase the aircraft range and manoeuvrability.  Although the flights were not of sufficient length to merit the prize in the competition organised by the Polytechnischen Gesellschaft, the commitee of judges decided, with the agreement of the Director of National Sporting

[1] *Düennebeil was Haessler's second choice for a pilot.  In his (Haessler's) writings, he rather ungraciously refers to the low power output of Düennebeil compared with professional cyclists.  Karl Duennebeil was twenty-seven years of age at the time of the flights and was a glider pilot with a 'C' certificate.*

Aviation, to award the competitors a prize of DM 3000.  In addition Air Marshall
Goering donated a further DM 3000 to be given to Haessler, Villinger, and
Düennebeil "in recognition of their efforts, and for the purpose of furthering
their work".

Prior to the flights at Rebstock, a total of fifty-five test take-offs had
been made with 'Mufli', although none of these had been sustained flights under
muscle power.  These first trials were carried out at Dessau.  Flights at
Rebstock were characterised by the method of take-off.  The aircraft was
tethered to the ground at the centre of the fuselage, and rubber cables were
tensioned between this location and a pivot point several metres in front of the
aircraft.  On release of the tether below the fuselage, the aircraft accelerated
and climbed to an altitude of about three metres.  The pilot then commenced
pedalling and the aircraft was thus able to fly horizontally over comparatively
large distances.  Calculations showed that the energy accumulated using the
tensioned cable was insufficient to maintain flights of the length experienced
by Düennebeil, and some proportion of the energy must have been contributed by
the pilot.  Sceptics remained unconvinced, and the use of what was a very large
elastic band to enable a take-off to be made was the subject of adverse comments.
However the flight trials were received very favourably by the steering committee
of the Polytechnische Gesellschaft, who decided to double the prize for the man-
powered flight competition, raising it to DM 10 000.

Following changes to the control of the aircraft, and with the aid of a more
powerful pilot, Hoffmann, (who had been tested by Ursinus at the Muscle Flight
Institute), flights of up to four hundred metres were made, and at a demonstration of his aircraft in connection with a lecture entitled *Sailplanes with
Auxiliary Lift*, given by Haessler at a meeting of the Technical Committee for
Sailplane Flight of the Lilienthal Gesellschaft, in Hamburg, a flight of 427
metres was made.  This flight was on 17th November, 1936.

Further flights were made in 1937, including a record 712 metres at Meiningen
on July 4th of that year, and in February 1938, after making approximately one
hundred and twenty flights, the aircraft was given to the Air Transport Museum
in Berlin, but did not survive the war.

## HELMUT HAESSLER'S H4

Haessler today remains interested in man-powered flight, and has designed a
successor to the Haessler-Villinger aircraft, the Haessler H4.  His performance
goal, which even now seems very ambitious, was based on two criteria, these being
his definitions of 'restricted' and 'unrestricted' flight.  The Haessler H4 was
designed to meet both these criteria.  The goal for restricted flight was a
duration of one hour at cruise power, flying at a height equal to the span,
(i.e. outside the beneficial ground effect region).  Take-off would be unassisted.
Haessler considers the lowest acceptable performance to be a take-off using the
power-driven ground wheel, the pilot using up his energy in a short flight of
about five minutes duration.  Unrestricted flight would impose much greater
demands on the structural soundness of the aircraft.  The stresses to be
expected would be similar to those experienced by gliders, and hence the weight
would be higher than that for a design which would only be subjected to flying
loads expected in restricted flight, (see also Chapter 17).

Haessler's proposed H4 the dimensions of which are given in Table 4, would

have a weight dependent upon the selected flight category.  For restricted
flight the minimum gross weight would be 110 kg.  For unrestricted flight,
where the pilot would carry a parachute and more comprehensive instrumentation
would be necessary, the all-up weight rises to 142 kg, with an upper limit of
162 kg.

These weights are substantially greater than that of the 'Mufli', but, as
shown in the table, more dramatic increases in size, and in particular the
very high aspect ratio, are characteristics which make the machine very similar
to contemporary designs.

Table 4.  Characteristics of Haessler's proposed successor
to 'Mufli', the H4

| | |
|---|---|
| Wing Span | 25 m |
| Wing Area | 16.75 m$^2$ |
| Wing Loading | 6.6 - 12 kg/m$^2$ |
| Aspect Ratio | 37.3 |
| Chord | 0.67 m |
| Glide Ratio | 1 : 44 |
| Fuselage Cross-section | 0.35 m$^2$ |
| Propeller Diameter | 1.70 m |
| Propeller Efficiency | 84% |
| Wing Section | NACA 65$_3$-618 |

In retrospect, the very low wing span, and the high wing loading of the
Haessler-Villinger 'Mufli' seem to deny the possibility that the machine could
ever have achieved prolonged man-powered flight, particularly with an unassisted
take-off.  However, it is very interesting to note that at least we are able to
compare certain features of this aircraft with corresponding features of
machines such as the Hatfield, Southampton, and Wisley aircraft.  It was only
a small step from a glider of the early 1930s, to a man-powered aircraft, but
the results of this step were to provide the basis for most, if not all, of
the more successful man-powered aircraft to date.

## HANS SEEHASE

Hans Seehase, a contemporary of Haessler and, like him, an engineer, was
also drawn to the design of a man-powered aircraft for entry in the Frankfurt
Polytechnische Gesellschaft Competition.

Seehase had been concerned with the development of man-carrying kites for
several years prior to his involvement in man-powered flight.  He made use of
the construction knowledge gained in this rather specialised activity in
several features of the machine, illustrated in *Fig. 60*, but most obviously in
the layout of the wing-truss structure.

The machine had a wing span of 13 metres, and an empty weight of 36 kg.
There was a very short streamlined cockpit section (the pilot held in a
cycling position), and the transmission system used long cranks as well as

chains for the final drive, shown in *Fig. 61.* The structure of the wing was
based on the use of two duralumin alloy tubes as spars. Four ribs per side
were attached to these spars, and the whole frame was covered with doped silk.

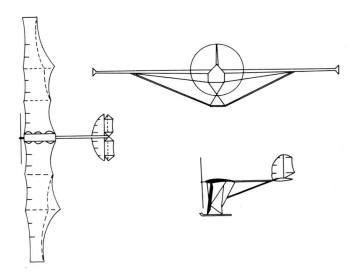

*Fig. 60*    The Seehase aircraft - wing loading was only
1 kg/m$^2$.

Small nose ribs were equally spaced in pairs between the main profile formers,
and to maintain sufficient spanwise rigidity, struts were incorporated, running
from the bottom of the fuselage to points at the middle of each semi-span.

*Fig. 61*    Transmission system of the Seehase aircraft

Although the wing structure of the Seehase aircraft had been hailed by some as offering substantial weight savings, this is a rather doubtful premise, and the aerodynamic efficiency of the wing was probably much less than that of for example, the Haessler-Villinger glider-type profile[1].

A more remarkable feature of the aircraft was undoubtedly the propeller. The tractor unit had a diameter of 2.86 metres, with two blades. The blades, however, did not, as in conventional propellers, extend from the hub, they were attached to the ends of constant chord aerofoil section struts of slightly under one metre in length. This gave a blade length of 45 cm, the overall effect being paddle-like. With blades of lime wood, and a rotational speed of 300 rpm, Seehase claimed rig thrusts of the order of 7.5 kg. How reproducible these results were when applied in the prototype aircraft remains in doubt.

## A GERMAN MISCELLANY

Whilst Haessler's machine was without doubt the most successful of its type in Germany in the 1930s, numerous other man-powered aircraft were designed there and several of these were constructed. Some of these have been described previously.

John Kaltenecker, over the period 1931-36, designed and operated a flapping wing aircraft which was catapult-launched. The layout was conventional, with a wing span of about seven metres, but a very large tailplane was provided. The flapping motion was imparted directly through the pilot's arms.

In 1935 August Fischer constructed a solidly built part ornithopter and part fixed wing aircraft. The pilot was positioned forward of the main wing, and was intended to control the aircraft using a small canard wing at his feet; tailplane and rudder were also provided. A tricycle under-carriage was used to ensure stability on the ground, and by pedalling the pilot was able to rotate the wing tips via shafts running immediately below the fixed wing sections. No take-off was reported, although the fixed wings could conceivably have enabled towed flights to be made.

What must have been the largest man-powered aircraft of the decade was that built by Englebert Zaschka, of Berlin, in 1934. With a tractor propeller having a diameter of 2.44 metres, and a wing span approaching 15 metres, it came close to what we recognise today as a good design. However no attempt was made to streamline the fuselage, and the pilot sat in the midst of a steel and bracing wire framework of frightening complexity. Zaschka's machine was designed for unassisted take-off, and as such did not meet with success.

The *Berliner Tageblatt* reported at the end of 1935 that a man named Kaiser had made a flight using muscle power in Bohemia. This was claimed by the reporter to be the second ever successful man-powered flight[2]. The route of Kaiser's flight took him twice round the town of Plauschnitz, which was at

---

[1] *The Perkins Inflatable Wing, used in a man-powered aircraft of the early 1960s (see Chapter 10), may be regarded in a similar way as the Seehase wing. It also offered weight advantages, at the expense of aerodynamic penalties. The Perkins wing was much more advanced, for its time, than the Seehase type.*

that time rather more than a village.  As if this distance was not sufficient
Kaiser is reputed to have then set off over a wood where a wing snapped and
he crashed but escaped injury.  This claim is needless to say unsubstantiated.

* * *

---

[2] *The first flight, or more correctly, aircraft to fly, in this context would
be that of the Haessler-Villinger aircraft, which first flew on August 29th,
1935.*

*Fig. 49.* Brustmann's ornithopter glider following
a towed launch at Wasserkuppe in 1929.

*Fig. 55.* An early photograph of the Haessler-Villinger
'Mufli'.

*Fig. 56.* Haessler at the controls of his aircraft,
1935-36.

*Fig. 57.* 'Means of entry!  The Haessler-Villinger aircraft in 1937.

*Fig. 58.* Construction of the Haessler-Villinger
man-powered aircraft, 1934-35.

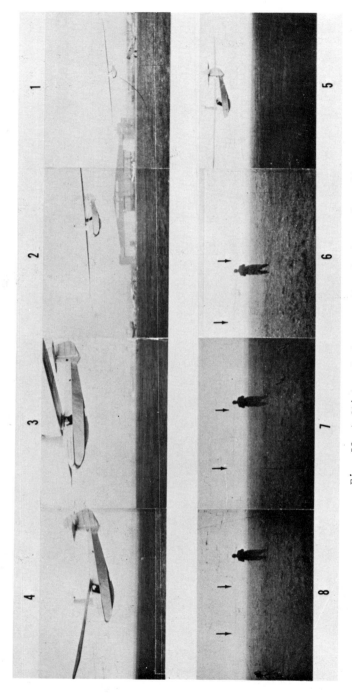

*Fig. 59.* A flight of the Haessler-Villinger 'Mufli', showing catapult-assisted take-off.

*6*

# The Spotlight Turns to Italy and Russia

# 6

In the mid-1930s the limelight quickly moved from Germany to Italy, where Enea Bossi was engaged in the construction of an advanced man-powered aircraft. Russian scientists were consolidating their position as experts on ornithopter technology, and experiments, directed at furthering their knowledge as applied to man-powered flight, were under way. Equally dramatic but less valuable contributions were noted in Britain, and a few minor attempts at defeating gravity using muscle were reported in the world's press in this period.

## ITALY IN THE 1930s

Enea Bossi was born in Italy and attended college in Milan at the beginning of the twentieth century. His interest in aviation was encouraged by the successes of the Wright brothers in America, and continued to grow although, or perhaps because, many of his contemporaries remained sceptical about the possibility even of powered flight. After graduating in physics and mathematics, he commenced the design of his first aircraft, a glider which could also carry a petrol engine.

Bossi's father was a far-sighted man who did not share the general disbelief in the possible performances of powered aircraft, and in 1908 he gave his son sufficient money to enable the aircraft to be constructed. It was completed in 1909 and was tested in its glider configuration, towed by a car. After its first successful flight, Enea Bossi took a friend for a trip, and on this second flight he attempted a turn at an altitude of 15 metres. In his recollections of this flight, he tells how his friend became frightened, and grabbed his arm. The aircraft crashed, luckily with no serious consequence to the two fliers, having been airborn for a total time of thirty seconds!

Not letting this setback deter his enthusiasm for flight, Bossi set about the reconstruction of his aircraft in 1909, this time using a 21 kW engine for propulsion. After a few months of test flying he was able to carry out flights of a few kilometres and perform simple manoeuvres. Such was his success that he was kept fully occupied designing and manufacturing sea-planes for the Italian Navy.

Bossi considered that his interest at the time in powered aviation was on a parallel to our current interest in man-powered flight. As very few people believed at the beginning of this century that man could fly using mechanical engines so in the 1920s and 1930s they were sceptical of man-powered flight. Initially, only short flights would be made, but as aeronautical expertise developed in the field of aerodynamics and materials, the concept of man-powered flight would become a familiar reality, depending solely on scientific progress in adapting the power available in the most efficient way.

Bossi became interested in man-powered flight in 1935, after reading that

an aircraft powered by a 0.76 kW engine had, it was claimed, successfully
flown.  Calculations made by Bossi to satisfy his curiosity about the minimum
power required to fly an aircraft showed that such a flight, with a power
requirement of the order of 0.7 - 0.8 kW, would be feasible.

During this period his work involved a considerable amount of travel, both
in Europe and across the Atlantic, and Bossi took the opportunity to use
travelling time to the best advantage by considering the problems associated
with the design.  As Bossi did not have access to information on the power
developed by a man cycling, mainly because little work had been done up to this
period, he had to obtain by his own calculations and experiments the thrust of
propellers with various geometries, (Bossi was an expert on propeller design)
and also determine the optimum pedal revolutions for power outputs extending
over a period of several minutes.  In addition he carried out experiments to
prove the feasibility of man-powered flight.

While Bossi was working on his aircraft, the Italian Government announced
a prize of approximately £2000 for a flight of one kilometre using man-power.
Bossi was at this time an American citizen, having left Italy shortly after the
end of the First World War, and he was disappointed to find that only Italian
citizens were eligible to enter this competition.  Having travelled to Rome to
file his entry, Bossi did not discontinue his work, but arranged for newspaper
representatives to witness any flights of the aircraft.  He had previously had
discussions with Vittorio Bonomi, a friend who was to be responsible for most
of the construction work on the aircraft.

The first trials in his experimental programme utilised a cyclist towing a
glider.  A tow rope with a scale mounted in front of the glider pilot, reading
from 0 - 14 kg, was attached to a bicycle ridden by a professional cyclist.
The glider pilot was to watch the scale and also a speedometer connected to one
of the cycle wheels, indicating the speed at which the glider took off.  Tests
were conducted on a quiet road approaching an airfield near Philadelphia, in
the United States.  Whenever the pilot considered that his speed was sufficient
for take-off, the glider was lifted about one metre and then brought back to
earth.  The tests indicated the pull that the cyclist would produce, the bicycle
being geared to operate at a speed no greater than 40 km/h.

The second test, run in Paris, was carried out by replacing the chain drive
to the rear wheel of the bike by a drive to a pusher propeller mounted on the
cycle frame behind the cyclist.  Bossi himself achieved a speed of 32 km/h using
this form of propulsion, while his assistant managed 37 km/h.  The gyroscopic
effect of the propeller was great enough to upset the stability of the platform
at these speeds and it was decided to carry out subsequent measurements on a
static test rig, *(Fig. 62)*.  It also indicated to Bossi that two contra-rotating
propellers would be required to eliminate this instability, although of course
he was misled in assuming that the aircraft would have similar inertia
characteristics to the cycle.

## THE BOSSI-BONOMI AIRCRAFT

Bossi concluded from these tests that man-powered flight was possible
provided that sufficient care was taken in the construction to minimise the
weight.  He concentrated next on the selection of an aerofoil section for the
main wing.  The choice, made from a short-list of three, the NACA F12, NACA
23012 and Göttingen 652, was based on cruise power requirements for flight, and

the first of the three, the NACA F12 (or NACA 0012-F1) was adopted, the cruise
power being 0.67 kW, a very high figure when compared with the aircraft designed
during the 1960s.  Further work on propeller design was then done, Bossi finding
it necessary to try a number of different pitches and diameters.  The wing area
and weight required for flight, based in part on Bossi's experimental results
on speed and power output, were also the subject of a large number of calculations, basically a crude optimisation procedure, aimed at obtaining the best
combination of all the variables.

A choice between a biplane or monoplane configuration now had to be made.
Bossi thought that there was no great advantage of one over the other, but
weight was the primary influence on his adoption of a monoplane.  He called his
aircraft the 'Pedaliante', which, literally translated from the Italian means
'fly by pedals'.

Bonomi, a sailplane manufacturer, considered that Bossi's design was
feasible and had a fair chance of success.  As a result, Bonomi was given the
contract to build the aircraft at his works near Milan, with a specified weight
of 81.7 kg, including a 9.1 kg contingency.  The safety factor was to be 2.0.
However, the Italian airworthiness requirements were rigorously applied to the
'Pedaliante', in spite of the fact that flying at such low speeds close to the
ground was unlikely to result in a serious accident.  Consequently increases in
strength meant that the empty weight of the aircraft reached almost 100 kg.
Bossi has often stated that his aircraft was 30 kg overweight and the vast
majority of this was due to the airworthiness requirements.

The wing span was 17.7 metres, and the aspect ratio of 13.4 was one of the
lowest ever seriously considered for a man-powered aircraft.  (This compares
with 18.8 for the Haessler-Villinger aircraft).  Most of the structure was of
wood, as illustrated in *Fig. 63*, and a conventional wing structural layout was
adopted - stiff leading edge and a strong main spar.  The fuselage consisted of
a triangular frame linking wing, tail and pilot position, and the fairing was a
light wood structure covered with fabric.  Two tractor propellers were driven
through bevel gears, power being transmitted from the chain-wheel through a
shaft.  The chain-wheel was also connected to the under-carriage wheel, the
ground-wheel being geared up to produce a ground speed of 40 km/h when
pedalling at 120 rev/min, the maximum design rating.

A variety of propellers were tried, their diameters ranging between 1.9
metres and 2.25 metres.  The propellers were laminated using balsawood, with a
hardwood core which also formed the small hub, *(Fig. 64)*.

The control surfaces were normal, apart from the controls on the main wing.
The wing profile used was characterised by a very thin trailing edge which took
up about fifteen per cent of the chord.  At first Bossi used spoilers for
lateral control, but it is believed that at a later stage ailerons were added,
these being mounted above the upper surface of the wing.  At this time spoilers
were a somewhat revolutionary means of control, and were untried even on gliders;
thus a considerable amount of time was spent familiarising both the pilot and
the designer with their behaviour, and minor modifications were made as a result
of experience gained during taxiing trials.  During these trials two additional
wheels were added to stabilise the aircraft, similar to the wheels attached to
some bicycles to assist a child learning the rudiments of cycling.  Three views
of the aircraft are shown in *Fig. 65*.

After a number of other minor modifications the aircraft was ready to fly,

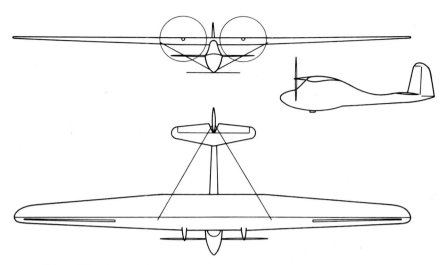

*Fig. 65*  Three views of the Bossi-Bonomi 'Pedaliante'

and the pilot, Emilio Casco, made the first flight in 1936.  Following a few
practice runs, the aircraft took off under its own power and flew a distance
of 91.4 metres[1].  Casco was, of necessity, a particularly strong man, for the
minimum power required for flight was calculated to be almost 0.7 kW, and it
was possible for him to develop 1.15 kW for short periods.  Sherwin[2] has
recently shown that this flight was entirely feasible, being within the pilot's
capabilities.

Bossi was particularly pleased with the results of this first attempt, and
considered that with further training it would be possible to achieve flights
of the order of 1.5 km.  However, it was found difficult in later tests to even
achieve the 91.4 metres flight length, and it was realised that atmospheric
conditions played a singularly significant part in the success or otherwise of
any take-off attempt.

However, further successful flights were made up to the end of 1936, the
longest being approximately one kilometre.  This distance was flown at Cinisello
airfield on 23rd December of that year, and it is important to note that two
right-angle turns were made during this flight.  The launch in this case was
assisted by a form of catapult, and airspeeds of up to 50 km/h were recorded.

Bossi, writing in 1960, states that over forty-three attempts to fly were
made, only half of which were successful, including the maximum distance
flight above.  Shenstone, an authority on man-powered flight, had discussions

[1] *Helmut Haessler, in a letter to 'Flight International' and in correspondence
with the author, states that the Bossi-Bonomi aircraft never took off unaided.
He also concludes that this machine was incapable of horizontal sustained
man-powered flight.  The flights mentioned by Bossi are considered by him to
be merely extended glides.*

[2] *K. Sherwin.  'Man-Powered Flight'. M.A.P., Hemel Hempstead, 1971.*

with Bossi in 1960, and in his writings he mentions that eighty flights were made by the Bossi-Bonomi aircraft.  However, it appears that Bossi's figure of forty-three refers to attempts made without the assistance of a catapult or stored energy at take-off.  *Figures 66 and 67* show the aircraft in flight.

Shenstone, in notes on a meeting he had with Enea Bossi in New York on 5th April, 1960, states:

"Mr. Bossi informed me that his 'Pedaliante' was flown about eighty times, forty take-offs being entirely by the pilot's efforts, unaided by energy storage.  Take-off speeds were of the order of 20 - 25 miles/h (32 - 40 km/h).  (Take-off) runs off concrete were 135 - 180 metres.  At 100 yards (90 metres) the aircraft began to lift.  It was essential to keep the aircraft at low incidence during a run, permitting the aircraft to fly off rather than be pulled off.  Even so the take-off was marginal, on some days not being possible.  The best weather for take-off was when it was cool and humid.  The maximum heights reached were about 30 feet (9 metres).

Bossi stressed the main problems as being the weight (the 'Pedaliante' was 67 pounds (30 kg) overweight) and propeller design.  The low wing loading chosen was not found to be too low ..."

Efforts were made to incorporate an energy accumulator to assist take-off, one being based on the use of a rubber band type device which was tensioned by the pilot before the flight.  Several technical difficulties were never overcome, however, and stored energy was never able to contribute to the pilot's efforts.  The tests had to be terminated because Bossi was unwilling to invest more money in the project, having spent approximately £5000, much more than he had originally anticipated[1].

Some considerable time after his work on 'Pedaliante', Bossi reviewed the progress he had made and the difficulties encountered, and he was able to make a number of useful recommendations to other designers of man-powered aircraft.

He was a great advocate of very large propellers for man-powered aircraft. The maximum diameter he reached on his machine was only limited by the fact that any further increase would have brought the tips of the propellers into contact with the fuselage.  Bossi found that every increase in diameter he made resulted in a noticeable gain in thrust.  (The propellers used were all laminated, based on a central backbone of strong wood, which also formed the hub).  The weight penalty imposed by the Italian airworthiness regulations was probably the greatest hinderance to long flights.  Bossi has in fact stated that had his aircraft been built to the original specification, in which the empty weight was 72.6 kg, flights of up to six kilometres would most likely have been possible.  With materials in existance today, his aircraft could be built with a weight considerably less than even this figure.  His estimated power requirements, however, remain very high when compared with current machines.

Most of Bossi's recommendations to designers were incorporated in later aircraft *(Fig. 68)*, although the results of various design teams seem to refute

---

[1] *Villinger suggested a competition or comparative trial for the 'Mufli' and 'Pedaliante', to be held in Germany.  The offer was not taken up, and Shenstone suggests that this may have been due to pressure from the Italian government.*

one statement that Bossi stresses several times in his writings on the subject,
namely the necessity for two contra-rotating propellers. His adoption of this
design feature was, as mentioned above, brought about by his earlier experiences
with a single propeller mounted on the bicycle. Apparently, the gyroscopic
effects were very noticeable, and although it is quite conceivable that this
would be the case on a laterally unstable machine such as a bicycle, the
effect on an aircraft is much reduced owing to the inertia effects.

## SERIOUS RUSSIAN EXPERIMENTS

One of the leading glider designers in Russia, V.I. Cheranovski, was also
a great exponent of man-powered flight, and demonstrated his ideas through the
construction of a number of machines. The first of the series was built in
Moscow in 1921, and was a biplane with articulated wings. Opposite upper and
lower wings were interconnected and given their flapping motion by means of
pilot-operated pedals. The wings had a degree of span-wise flexibility, but
the configuration required an assisted take-off before any flapping thrust was
developed, and proved to be unsuccessful.

A second machine manufactured in 1931 was a flying wing ornithopter
weighing 100 kg. The propelling surfaces were concentrated at the wing tips
and a dorsal fin was added for stability. The apparatus was strapped to the
pilot using a special harness, and although originally the pilot had to bear
the whole weight of the machine, a skid was later provided. It was intended
that flights would be made following a towed launch, but the flapping amplitude
was very restricted, and the lack of experience with flying wing designs also
contributed to the failure to achieve flight. Gliders similar to this have
flown in Russia, however.

Cheranovski realised that he had introduced too many new concepts in one
machine, and wrote: "In drawing up projects of such new experimental machines
like the ornithopter, it is best to avoid simultaneous solution of other
problems". Between 1935 and 1937, Cheranovski made a third man-powered aircraft,
known as the BICH-18 *(Fig. 69)*. This machine was an ornithopter biplane-glider,
and the principle of operation resembled that of the 1921 model.

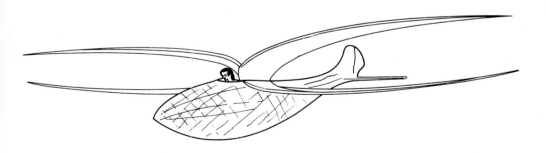

*Fig. 69* V.I. Cheranovski's ornithopter design.

With a combination of horizontal flight velocity and a vertical flapping velocity, an angle of incidence was produced which was large on the wing downstroke and small on the return stroke. The leverage was such that when the two upper wings were moving down, the lower pair were being raised. Cheranovski was able to alternate the lift increment between the upper and lower set of wings, resulting in a steady lift output for the system. By proper design of the wing incidence, a thrust increment would also be produced, this being necessary for a sustained flight, even with assisted take-off. The flapping of the wings was via leg muscles, the arms being reserved for control.

BICH-18 was constructed of plywood, and covered with a canvas material. The wing span was only 8 metres but the biplane configuration gave a total wing area of 9.1 m$^2$. With a gross weight of 130 kg, successful flights were made in a gliding mode (P.A. Pischechev was the pilot), but its performance as an ornithopter is unknown.

In parallel with Cheranovski's work, two Soviet aeromodellers constructed scale ornithopters. Trunchenkov and Stepchenko flew these during the period 1932 - 35, but no full scale prototypes were made.

*  *  *

At the Osoaviakhim gliding centre, an organisation dealing with Russian strategic industrial policies, Smirnov's man-powered ornithopter was tested in 1936. The two wings, with bamboo framework, were covered with a silk cloth and attached to special arm bars with wing beat limiters. The pilot was suspended by a roller pulley from an overhead cable 300 metres long and he was thus able to slide downhill flapping his wings. Fabric stretched between the airman's legs acted as fin and rudder. Chekalin, who 'flew' the aircraft, made a total of thirteen descents on the wire, and it is claimed that he relieved the tension in the cables during the run. Stability and control were said to be good, although no reports are available on attempts made without the assistance of a wire.

Another Russian design had fixed wings with flapping tips, the movable sections being 3 metres in length. The aircraft was catapult launched, and as a result of the flapping of the outer wings, caused by the muscular efforts of the pilot, there were increases in flight distance of 30 per cent, and in endurance of 58 per cent compared with an equivalent conventional glider *(Fig. 70)*.

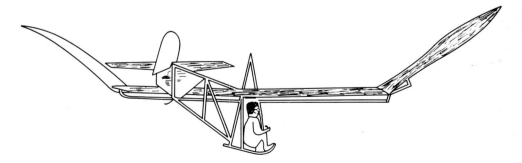

*Fig. 70* A glider with a flapping flexible wing, believed to have flown in Russia.

An insight into the limitations of the man-powered ornithopter as envisaged by most of the designers may be gained by the following conclusions of a Russian ornithologist of the period (confirmed by experiments carried out by Savel'ev and Vinogradov)[1]:

(i)   The elimination of the gaps and the slot mechanism of the primaries (large wing feathers) made the wings ineffective in that the bird was compelled to adopt another flapping mode which was inefficient and exhausting.

(ii)  Glueing over the underside of the wings changed the nature of the flow past them and reduced the lift below that necessary for level flight.

(iii) Joining of all the feathers with glue, combined with the blocking of all gaps between the feathers, such that the wings were as those of a glider or aircraft, resulted in an inability to fly by flapping.

## JOHN D. BATTEN AND THE SEAPLANE

Batten wrote a thesis[2] in 1927 on the subject of his researches into man-powered flight, which must have involved a vast amount of thought and experimentation. For the period, however, his line of attack appears a little dated.

He studied in detail the wing forms and musculature of a number of birds, rightly laying considerable emphasis on the importance of the pectoral muscles, in particular the *pectoralis major*, which in turn actuates the *humerus*, the first bone in the wing structure. The *pectoralis major* is primarily to depress the wing, while a second pectoral muscle, the *pectoralis secundus*, has the role of elevator for the wing. Batten obtained help from his brother in dissecting a pigeon, discovering that the *pectoralis secundus* weighed only three-sixteenths of the *pectoralis major* in this particular bird.

In making a distinction between the activities of various muscles, Batten introduced the term 'antigravity muscles' to describe those whose main function was to preserve the posture, or rigidity of a body, constantly resisting the influence of gravity. He suggested that the greater pectoral in a bird was an 'antigravity muscle', and that during gliding movements, exemplified by the larger birds, its activity was postural and it had an almost indefinite immunity to fatigue which is general in postural activities. He extended his thesis by suggesting that although the obvious function of the greater pectoral muscle was to create wing down-beats, by far its most important function was to prevent the wing flapping upwards, this being a postural activity analogous to the front thigh muscle in the human leg, whose main function is to prevent the knee from bending.

The basic aircraft design philosophy thus adopted by Batten, which is a logical extension to his above arguments, was to replace by a spring or some

---

[1] *This also confirmed the results of the work carried out near the beginning of the century by Dr E.P. Smirnov  (see Chapter 4)*

[2] *John D. Batten.   'An Approach to Winged Flight'. Dolphin Press, 1928*

similar device the postural activity of the pectoral muscles, and to use the
muscles in the man's arms to operate the transient motion of flapping the wing.
Although the arm muscles are comparatively small, it was believed that once
wing support had been achieved using other than man-power, wing motion could be
readily produced with little power requirement.

In 1925 a conveniently noteworthy glider flight was made by a M. Auger near
Cherbourg. The magazine *Flight* reported that Auger ascended to a height of
720 metres, returning to land near his point of take-off. This was regarded
by Batten as a 'flight with locked wings'; Batten surmised that if the wings
had been 'unlocked', no doubt they would have folded upwards with disasterous
results. If, however, they had been restrained by springs strong enough to
hold the wings horizontal until an additional load was applied, the ability of
the pilot to exert this additional load, which Batten believed was small, would
be a demonstration of actual 'man-powered flight'.

In order to implement this, Batten required a suitable spring system, and
an experimental rig which "shall not involve too great a risk to life and limb".
His plan for satisfying the first of these requirements was based on the use
of a 'catapult torse', this being a belt mounted between two stays and twisted
with a central lever. Silk cord was proposed for the belt material, because
of its superior tensile strength when woven.

The first machine designed was so arranged as to utilise as much of the
pilot's muscle power as possible in flapping the wings downwards. The pilot
stood between the two wing roots, to which were attached bars which he grasped.
By bending knees and arms he could apply considerable effort in moving the
wings. Two sets of torses actuated two main wing spars and a working model was
made to demonstrate the operation. The torses were so arranged that in moving
the wing tips down, the tension in them was eased, and in moving the tips
upwards, i.e. folding the wings, the tension increased, restricting movement
and hence 'locking', avoiding wing failure.

Apart from rigs to measure the performance of silk torses, no full scale
prototypes were constructed. However, a man-powered seaplane was proposed,
which is believed to be unique. Batten appreciated the need to flap in such a
way as to produce forward thrust, and realised that the effort for propulsion
would arise solely from the pilot's efforts.

In the postscript to his treatise on man-powered flight, Batten remains
optimistic, in spite of the forecasts of some of his contemporaries, and the
contraptions produced by them, two of which are illustrated in *Figs. 71 and 72.*

"I have been told: 'The usual calculations made lead to the conclusion
that man cannot fly by his own muscular energy' ... But as birds undoubtedly
fly by their own muscular energy, such an opinion needs the support of some
argument based upon an irremovable difference between men and birds ... It
is in muscular equipment that birds have a conspicuous advantage over men.
The whole of my invention is ... supplementing man's deficiency in this
respect ..."

He concludes: "It would, moreover, be no small boon to humanity that flight
should be silent".

                                        John D. Batten

Kew Gardens, Surrey.
December, 1927.

## CLEM SOHN

A rather dramatic attempt at man-powered flight took place in Southern England in the Spring of 1936.

Clem Sohn, an American, made full use of powered aircraft to climb to an altitude of 2750 metres, flying as a passenger in a B.A. Swallow piloted by a Flt. Lt. J.B. Wilson.  Sohn carried two bat-like wings which were actuated by his arm movements, and a tail which was supposed to act as a rudder, direction being dictated by leg motion.  Rather prudently, he also carried two parachutes.

The popular press of the day had apparently given great publicity to the possible achievements of Sohn, even suggesting that an eight kilometre glide could well be achieved if the expected performance was obtained by flapping the wings.  As a result the event attracted a large crowd of spectators, the more knowledgeable of whom no doubt remained stoically sceptical regarding the whole escapade.

As planned, Sohn made his exit from the Swallow at 2750 metres.  Instead of gliding however, an almost vertical descent resulted, clearly visible as he was carrying a smoke-producing cannister, with a few slight deviations from a constant path which could be attributed to arm movements.  After a quite spectacular dive of 2450 metres, Sohn opened one of his parachutes and glided to earth.

The magazine *Flight* commented at the time that:

"it will be a magnificent stunt for air circuses, but all this talk of seriously developing it (and even of carrying 'compact and light batteries' to actuate wing-flapping mechanism) merely suggests a return to the dark ages before the dawn of aeronautics;  as Francis Bacon wrote: 'The Leucacians in ancient time did use to precipitate a man from a high cliffe into the sea ... fixing into his body divers feathers, spread, to breake the fall'."

Although uninspiring as an attempt at man-powered flight, Sohn's escapade must be one of the earliest intentional (or semi-intentional) free flight parachute drops, which are now so popular at displays throughout the world.

* * *

Other minor attempts recorded before the start of World War II include the following:

The Borghese-Parizza Cyclo' Voilier, made in 1931, was a muscle-powered ornithopter tricycle.  The apparatus was tested in Paris during the 1930s, and failed to leave the ground.

* * *

The Fiorentini ornithopter consisted of two flapping wings and a framework. It weighed 7.27 kg and had a span of 4 metres.  This reduced model was tested in 1938 at Borgaretto, and was capable of flying for a few seconds.  A flapping frequency of 10 - 12 beats per second was achieved, beyond that likely to be attained by the human arm.

*  *  *

    W.H. Herman, of San José, California, constructed a man-powered ornithopter following twenty years of experimentation.  The machine was strapped to the operator's body and the wing motion was produced by means of hand levers. The pilot wore a cap which, through cables, actuated elevators in the rear. Nodding of the head produced control.  Take-off was assisted;  a trailer carrying the pilot and his apparatus was driven up to take-off speed.  The ornithopter weighed 25.8 kg and had a wing spread of 11 metres.  Wood and wire bracing formed the frame, which was covered with light canvas.  No success was achieved.

*  *  *

    In 1934, Lindmann constructed an ornithopter with flapping wing tips, mounted on a bicycle.

*  *  *

    The Ricci ornithopter consisted of two flapping wings hinged to a central fuselage.  Only a scale model was built and tested, but Ettore Ricci spent a number of years in his native Italy studying unconventional forms of flight.

*  *  *

*Fig. 62.* Testing the propeller of the Bossi-Bonomi
aircraft in Paris, 1935.

*Fig. 63.* Structural detail of the Bossi-Bonomi
aircraft is open to view in this early photograph.

Fig. 64. The Bossi-Bonomi aircraft in 1936. The propellers were a noteworthy feature of the design.

Fig. 66. The Bossi-Bonomi machine in flight, in front of an impressive scene.

*Fig. 67.* Another view of 'Pedaliante'.

*Fig. 72.* A successor to the early 'Aviettes' –
Francois Baudet's aircraft at a display for
the press.  No flights were made.

*Fig. 71.* Alois Santa, the French designer, in his man-
powered aircraft, 1923.

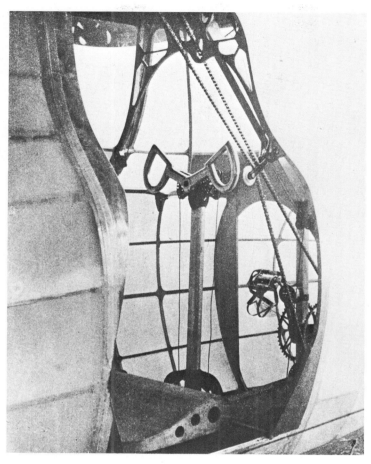

*Fig. 68.* The straightforward primary drive and control
column of the Bossi-Bonomi aircraft are evident in this photograph of the cockpit.

# 7

# A
# Period
# of
# Reappraisal

*— Measurements of Man-Power and Studies of Bird and Insect Flight*

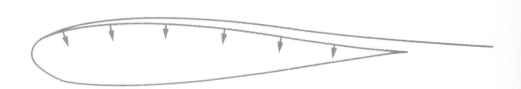

# 7

The Second World War interfered temporarily, with one minor exception[1], with the further development of man-powered aircraft, but the general progress in aeronautical science was, as usual, accelerated by the requirements for more advanced aircraft, and by 1945 the aerodynamics and structural design methods had undergone a minor revolution.  The monoplane was now shown to be totally superior to the biplane[2].  Engine efficiency in terms of power per unit weight had greatly improved, and the reliability of aircraft had increased at a similar rate to the growing sophistication, not least with the advent of the jet engine. Some countries, affected by the shortage of raw materials, had earnestly begun the development of man-made materials such as synthetic rubber, later to be followed by the vast range of plastics and organic chemical compounds, many of which offered increased strength to weight ratios over their more natural competitors.

Although the above developments were obviously aimed at larger and faster powered aircraft, inevitable advantages would be seen by utilising the techniques in light aircraft construction.  The age of special alloys and synthetic materials which was just beginning was to play a large part in the later establishment of man-powered flight as a small but seriously supported branch of aviation.

Between the advent of World War II and the generation of interest in man-powered flight at the College of Aeronautics, Cranfield in the mid-1950s, some comment and research activity, widely scattered from a geographical point of view, was evident.

The decade immediately following the Second World War was, however, rather lean as far as man-powered flights were directly concerned.  This must partially be attributed to the fact that the two major pre-war attempts, in Germany by Helmut Haessler and in Italy by Enea Bossi, had, in common with numerous other ventures, been influenced by political and military pressures prior to 1939, and neither country had time after the war for what then appeared a rather frivolous exercise.

No definite direction was noticeable in the work, and this was partly as a result of the fact that no co-ordinating body had been formed, or had survived the war.  Also, those starting to work on man-powered flight in South America, the United States, Britain and Europe were probably unaware of the activities elsewhere until the results of their deliberations had been published in the

---

[1] *In 1944 Bedford D. Maule of Ohio successfully flew, after assisted take-off, an ornithopter-glider, wing movement was restricted to an arc of one metre.*

[2] *The biplane was not necessarily eclipsed by the monoplane as far as man-powered flight was concerned.  The author carried on a considerable amount of research on this topic and his conclusions are given in Chapter 15.*

learned journals.

## POST-WAR REVIVAL OF INTEREST

In the late 1940s occasional evidence of interest in man-powered flight was seen in Britain. Although at this time the official organs of the technical societies concerned with aviation in Britain did not include articles on man-powered flight, a few of the aeronautical journals whose appeal was directed mainly at the informed layman did carry reviews of the work carried out in Italy and Germany in the 1930s, and several proposals were made for improving the design of glider-type aircraft so that power requirements would come within the range of that produced by a man.

Perhaps the most considered analysis of the problems of man-powered flight was published in 1948 in the aviation magazine *Aeronautics*. The author, Brian Worley, used the power measurements of Ursinus as his basis for the design. (These measurements are discussed in Chapter 5). His calculations on the powers required to fly the Bossi-Bonomi and Haessler-Villinger aircraft, detailed in *Fig. 73*, show the strong influence of airspeed on the power requirement. The two horizontal axes, representing peak and cruise powers, are calculated assuming a 75 per cent propulsive efficiency.

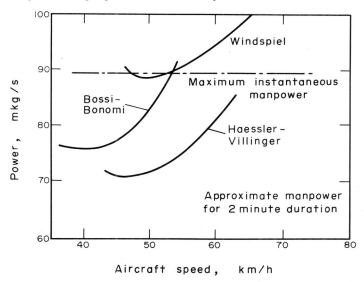

*Fig. 73* Power requirements of the Windspiel glider, compared with the 'Mufli' and 'Pedaliante' man-powered aircraft.

A curve showing the power requirements of the Windspiel D-28b glider, at that time the lightest performance sail plane built is included, permitting an interesting comparison. The glider had a wing span of 12 metres, a wing area of 11.4 m$^2$ and an all-up weight of 152 kg.

Worley's derivation of the equations for expressing the required power in

terms of the main variables available, showed that low weight was of prime
importance, preferably in conjunction with a large wing aspect ratio, in order
to minimise the induced drag.  Of secondary importance were the use of a low
wing loading, and the advantage obtained by flying at a low value of the lift
coefficient.  He made the following suggestions for achieving these aims:

   (i)   Reduce the strength of the aircraft.
  (ii)   Determine the optimum wing area.
 (iii)   Determine optimum span.
  (iv)   Improve the drive mechanism.
   (v)   Fly close to the ground.
  (vi)   Improve the wing section design.

Using the results obtained by Ursinus, Worley was able to give an indication
of the maximum allowable power for a man-powered aircraft.  With these figures
he then determined the major parameters of an aircraft which would come within
these power limits.  It is interesting to briefly review to what extent his
recommendations, given above, have been implemented, as a result of independent
research.  Certainly (i) the strength of the aircraft has been reduced, a
glider-type structure being more than adequate.  Flying close to the ground has
enabled this to be done, (v).  An optimum wing area (ii) and span (iii) are
difficult to define, although an infinite span is ideal.  Our thoughts are
regimented to some extent by the Kremer Competition, and manoeuvrability imposes
restrictions on span (although some groups would disagree).  Considerable
research has been done to improve the drive efficiency (iv).  Attempts have
been made to adapt existing aerofoil sections to make them more attractive to
man-powered aircraft.  This is probably a field in which some gains could
usefully be made in the future decade.  In general Worley's perception has
proved accurate.

The aircraft thus derived from Worley's suggestions was a monoplane with
conventional propeller and pedalling mechanism, the latter unit weighing 9.1 kg.
With a structural weight of 33.1 kg and a crew weight of 63.3 kg, this gave an
all-up weight of 105.5 kg.  He assumed a wing area of 11.9 $m^2$, corresponding to
a wing loading of 8.8 kg/$m^2$.  Worley stressed that the parameters had been
calculated assuming the use of a high speed wing section.  At the time little
data was available on thick laminar flow profiles which could be operated
successfully at low speeds.  His foresight in emphasising the need for such
profiles was to be proved accurate ten years later when the first successful
flights were made in Britain.

## AUGUST RASPET AND 'BOUNDARY LAYER' THEORY

In the early 1950s, August Raspet, an aeronautical engineer working at
Mississippi State College on glider design, took one aspect of aerodynamics
- drag reduction - and applied it to man-powered flight.  He rejected the
ornithopter as a viable flying machine primarily because the insufficient
experimental and theoretical knowledge of the performance of flapping wings
made it almost impossible to predict the power requirements.  Raspet also
discounted the man-powered helicopter, using as a basis for his arguments the
results obtained by Gropp in Germany in 1936.  (Gropp had determined that an
aircraft with an all-up weight of 100 kg and a rotor 17 metres in diameter
would require 1 kW.  A pilot with an average weight of 75 kg would thus have
to be supported and propelled by a structure weighing only one quarter of this,
and even if such an efficient structure were available, the duration of flight

would have been limited to only twenty seconds.)

   Raspet thus chose a conventional fixed wing aircraft layout for his man-powered design, and was able to apply much of the pioneering work he had done in sailplane design to suggesting ways for reducing the drag, and hence the power, of this aircraft.  At this time the comprehensive results of Ursinus on the power output of man were still the most reliable available, and Raspet used the figures obtained for an athlete using both legs and arms, the arms being used to operate a hand crank.  The figures given were 1 kW for a period of twenty seconds, tailing off to a cruise power of 0.43 kW.  (It appears that both Worley and Raspet assume these high power outputs without considering the significance of the use of both arms and legs.  Both their proposals were for single seat aircraft, and no mention has been made of the system required for controlling the machine.  In fact Raspet states that the use of legs only by the pilot of the Bossi-Bonomi aircraft was a major deterrent to its success.)

   He appreciated the advantages of flying close to the ground, where reductions up to fifty per cent in the induced drag could occur, *(see Fig. 74).*

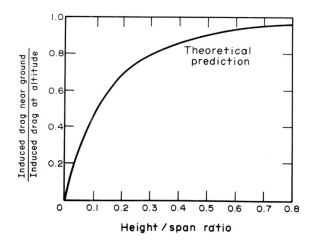

*Fig. 74* The variation of the induced drag as one approaches the ground.  (The drag in coefficient form is plotted against the altitude to span ratio, a convenient form independent of aircraft size).

However, not satisfied by the contribution possible by flying in ground effect, Raspet considered the features of soaring bird flight, where he found that the flow over the suction surface of a bird's wing was laminar over most, if not all, of the chord.  This was put forward as the only explanation for the extremely low drag of birds.

   On the assumption that the mechanism for maintaining low drag bird wings by preserving the laminar boundary layer were fully reproducable on a man-powered monoplane, Raspet suggested a machine with the following characteristics:-

| Span | 12.2 metres |
|------|-------------|
| Chord | 1.52 metres |
| Flying weight | 136 kg |
| Speed | 29 km/h (minimum) |
| Wetted area | 55.8 m$^2$ |
| Drag coefficient | 0.00465 |

The power required to fly such an aircraft at altitude would be 0.59 kW, reducing to 0.34 kW near the ground.

Raspet's work on reducing the profile drag of glider wings had been reported in 1951.  The technique used for maintaining the laminar flow over the surface is known as boundary layer suction, and Joseph Cornish also carried out similar work in America during the same period.  The basis of boundary layer suction is incorporated generally under the group of aerodynamic modifications which may be made to aircraft, known as boundary layer control. These techniques include blowing through slots, sucking and blowing through slits, and similar forced motion of air through perforations.  The components on which boundary layer control is mainly adopted are wings, where the aim is either to increase lift or decrease the drag.

There are two forms of boundary layer suction through perforations in an aircraft wing, the type proposed by Raspet and Cornish for man-powered aircraft. Laminar, or low drag boundary layer suction is used to prevent or delay the boundary layer transition from laminar to turbulent flow.  The second form, turbulent boundary layer control, is used to prevent flow separating from the wing, this occurring generally when the boundary layer has become turbulent. For example, this second type may be used to delay the stall of an aircraft wing at high incidences.

Cornish, assisted by Wells, both members of the Engineering Department at Mississippi State University, carried out an interesting exercise in which they applied the concept of boundary layer suction through a perforated wing to the Nonweiler aircraft.  (This machine is discussed in detail in Chapter 8).  They considered both laminar and turbulent boundary layer control, using results obtained in experiments carried out on a glider specially modified for the suction tests.

(The author carried out some work in 1965 on the possibility of adopting boundary layer control to man-powered aircraft.  It was found that although noticeable increases in the lift coefficient occurred, the suction being applied towards the trailing edge of the aerofoil under test, little significant drag reduction was evident.  One disturbing factor which would possibly limit the application to man-powered aircraft was that the suction caused distortion of the upper wing surface, as very thin plastic materials similar to those used on the Hatfield aircraft were used.  These distortions could prove to have unpredictable effects on the profile drag of the wing section, and a more rigid wing surface would be required.

At Glasgow University work on suction wings was also being carried out by undergraduates in Professor Nonweiler's Department.  Similar conclusions were reached, it being felt that they were not a useful feature.)

The drag of the aircraft may be divided into three major components, wing induced and profile drag, and the fuselage profile drag.  Boundary layer suction has little effect on the wing induced drag; however, it does reduce the wing

profile drag, *(Fig. 75)*.  There is a slight power penalty involved, owing to
the necessity to provide a pump to produce the suction.  This power may be
incorporated in the total drag as a 'suction drag', being a function of the
pressure rise of the boundary layer air and the suction flow rate.

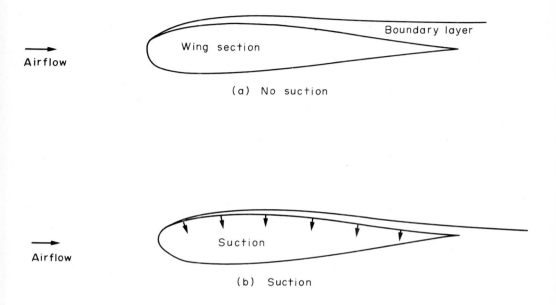

*Fig. 75*  A wing section with a normal boundary layer on the
upper surface (a), and with suction applied inside the wing
(b).  The boundary layer thickness is considerably reduced.

The effect of the application of boundary layer control on the power
requirements of the Nonweiler aircraft was determined by Cornish.  Including
in the power calculations an allowance for the suction drag, application of
laminar boundary layer suction reduced the power requirement by about 5 per
cent compared with that required for flight with an impervious wing.  Cornish
calculated that the blower and associated ductwork would weigh not more than
5.45 kg, or 2.5 per cent of the all-up weight.  Apparently the blower proposed
by Cornish was designed around its own power unit.  Any attempt to utilise
such a device in the aircraft entering the Kremer Competition, the international
event organised by the Royal Aeronautical Society, would necessitate the
adaption of the suction device to man-power.

## A CONTRIBUTION BY PHYSIOLOGISTS

The ability of man to fly aided by flapping wings attached to his arms was
the subject of many of the very early exercises on man-powered flight, but with
the advent of the theory of aerodynamics in the nineteenth century, and its
successful application to the fixed wing powered aircraft at the beginning of
the twentieth century, flapping motion became almost totally neglected as a

serious contender for producing successful flight, either by mechanical or muscular motivation.

However, in the mid-1950s two physiologists, Günther and Guerra, working at the Physiology Institute of the Medical School, Chile University, applied the theory of biological similarity to flight characteristics of a number of insects and birds.  They were able to correlate these characteristics, which included such features as wing beats per second, wing area and wing span, using a number of equations.  On extrapolation of the results to include the power and weights of a man, they concluded that a human could fly limited distances with small wings attached to his arms.

Using measured data for insect and bird flight obtained by Bjerknes, Günther and Guerra plotted the flight characteristics on a double logarithmic scale, and this is shown in *Fig. 76*.  The available data covered a range of creatures from the fly, with a body weight of 0.01 grams, to the stork, with a body weight of approximately 3500 grams.  Shown on this figure are four characteristics plotted as a function of body weight, these being the number of wing beats per second, the force of wing beat, the wing tip speed and the wing span.

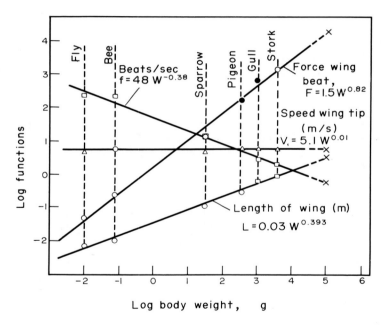

*Fig. 76* Flight characteristics of insects and birds as a function of body weight (logarithmic scales).

In Table 5 the equations correlating these variables and a number of others in terms of body weight are shown together with the limiting coefficients predicted by the theoretical biological similarity studies.  Variations in the limiting values were ascribed to the comparatively few sets of experimental data available, and their possible inaccuracy.  Further data obtained by Pütter in 1911 was plotted using the equations given in the Table, and compared with theory;  the agreement was found to be satisfactory.

Table 5.  Flight characteristics of birds and insects in relation to body weight (W).

| Functions | No. | Logarithmic equations: The theoretical limits are given in brackets | Body weight in | Correlation coeff. 'r' |
|---|---|---|---|---|
| Flight surface ($m^2$) | 1 | $\log S = 0.664 \log W -0.551$ (0.66) and (0.66) | kg | 0.995 |
| Pressure ($kg/m^2$) | 2 | $\log P = 0.337 \log W +0.55$ (0.33) and 0.00) | kg | 0.982 |
| Speed (m/sec) | 3 | $\log V = 0.168 \log W +0.923$ (0.16) and (0.00) | kg | 0.980 |
| Power ($W \times 10^{-1}$) | 4 | $\log P_o = 0.667 \log W - 0.555$ (1.16) and (0.66) | kg | 1.00 |
| Force wing beat (g-weight) | 5 | $\log F = 0.82 \log W + 0.185$ (1.00) and (0.66) | g | 1.00 |
| Speed wing tip (m/sec) | 6 | $\log V = 0.011 \log W + 0.709$ (0.16) and (0.00) | g | 0.378 |
| Length of wing (m) | 7 | $\log L = 0.393 \log W - 1.467$ (0.33) | g | 1.00 |
| Wing beats per second | 8 | $\log f = -0.382 \log W + 1.68$ (-0.16) and (-0.33) | g | -0.992 |

Using equation 4 in Table 5, it is possible to plot the power to fly as a function of body weight, data for birds ranging from the hummingbird, with a weight of 8 grams, to the condor weighing 12.7 kg.  These results show the quite large increase in power required with increasing weight (*Fig. 77*).

The application of these results to man-powered flight was simply carried out by assuming that a man with an average weight of 70 kg, and with an additional allowance of 30 kg for wings and other flight accessories, would come within the bounds of the correlations obtained for birds and insects. Substituting this all-up weight value of 100 kg into the equations in the table yielded the following data for the flight characteristics of a human being. These results may be alternatively obtained by extrapolation of the curves in *Fig. 77*, and the points corresponding to the assumed all-up weight are denoted by an asterisk.  The most noticeable fact shown by this work is the very low power requirement, of the order of 76 W, which is less than the power needed for brisk walking.  The only limiting factor likely to create difficulties was

| | |
|---|---|
| Flight surface | 6 m$^2$ |
| Pressure | 16.8 kg/m$^2$ |
| Speed | 18.2 m/sec |
| Power | 60 W |
| Wing beat force | 19.3 kg |
| Wing tip speed | 5.8 m/sec |
| Wing semi-span | 3.15 m |
| Wing beats per second | 0.59 |

stated by Günther and Guerra to be the air viscosity.  They assumed that three
main factors had prevented man from achieving his goal in this mode of flight.
The first was the lack of physiological data available of the type obtained by
them;  secondly adequate materials were not available for producing the wings
(although no suggestions are made as to the types of material which should be
used), and finally, little training had been carried out in previous attempts
to simulate bird wing motions.  The reason for the latter shortcoming was
probably due to the fact that man's eye was unable to perceive more than a
simple up and down flapping motion of bird wings, thus their attempts at
simulating the wing beating was restricted to this activity.  Slow motion
photography has revealed that the motion of the wings of birds is complex and
man would find the muscular and bone mechanisms necessary difficult to reproduce
in wood, alloy or any other materials.

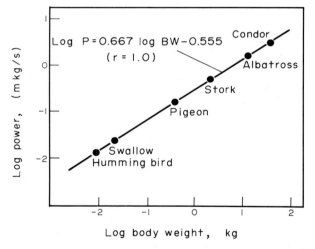

*Fig. 77*  Günther and Guerra's values for the power to fly
as a function of body weight.  An extrapolation to include
man is incorporated on the graph.

## THE 'ENTOMOPTER'

   Applications of largely untried theories to the solution of the problem of
man-powered flight were a common feature of most periods in the otherwise
logical development of the science in the twentieth century.  Some of these
theories were perfectly sound.  The attraction of man-powered flight as a forum

for the exploitation of these ideas was natural - the aircraft were comparatively cheap to manufacture and no vast research and development expenditure was required.  Any fairly talented amateur could 'have a go', (although this far from guaranteed success) and the necessary publicity, or in a few cases notoriety, was easy to obtain.

A short note in *The Aeroplane* of 1950 describes one such theory and the proposals of the American perpetrator, A.L. Jordanglou, for translating it into a man-powered aircraft.  His work, spanning two decades, was concerned with the 'Entomopter' (Greek 'entomon' - an insect, 'pteron' - a wing).  The author of the note referred to work by Professor A. Magnan in France, on the mode of flapping of an insect's wings, aided by high speed cine photography. Professor Magnan also developed a theory to describe hovering using flapping wings, and it was this latter work which attracted the attention of Jordanglou and his colleagues.  They constructed a number of models in an attempt to prove Magnan's theory.  Obviously the results obtained in these experiments were encouraging, because, writing in the journal *American Helicopter*, in 1949, Jordanglou proposed a man-powered aircraft based on utilising the lift obtained from flapping wings.  This 'entomopter' was to be flown by a single pilot, who would operate the wings via cranks, the motion of the pilot being akin to rowing.  The experimental studies of Jordanglou indicated one drawback, however - the rate at which the wings were required to operate was fifty strokes per second!

## BEVERLEY SHENSTONE

The man most instrumental in bringing the public's attention to man-powered flight, and in particular to enable them to recognise it as a serious activity requiring expert knowledge in aeronautical engineering, was B.S. Shenstone, then Chief Engineer of British European Airways[1].

His first two published notes on man-powered flight exhibited widely differing approaches to the problem.  An article reproduced in the *Sunday Times* of February 27th, 1955, concluded with the recommendation that further work on understanding the mechanism of bird flight should be done.  Seven months later in a lecture to the Toronto and Winnipeg Branches of the Canadian Aeronautical Institute (CAI), Shenstone suggested fixed wing man-powered aircraft, aerodynamically and structurally refined examples of the Haessler-Villinger and Bossi-Bonomi machines[2].

The newspaper article revealed some interesting facets concerning the performance of birds and fish.  Shenstone suggested that the speed at which they travelled, coupled with their musculature, could only be achieved if some form of boundary layer control were used.  He went on to recommend that a series of experiments be undertaken to ascertain the types of boundary layer

---

[1] *Beverley Shenstone was later to become a founder member of the Man-Powered Aircraft Committee, based at Cranfield, and the Man-Powered Aircraft Group of the Royal Aeronautical Society (Chapter 8).*

[2] *The lecture to the CAI consisted of a preliminary version of a paper presented to the Low Speed Aerodynamics Research Association (LSARA) in London on 12th November, 1955.  This paper, entitled 'The Problem of the Very Light-Weight Highly-Efficient Aeroplane', was reproduced in the CAI Journal during 1956.*

control used by birds and fish, and the ways for reproducing such mechanisms
in man-made machines.  The article stresses that the way to successful man-
powered flight is through the study of bird flight, following a much different
vein to that of his paper given before the CAI and LSARA.

(Recent studies, reported in *Nature*, (Vol. 49, p 234, November 5th, 1971)
carried out by Dr. M.W. Rosen and Dr. N.E. Cornford of the United States Naval
Undersea Research and Development Center at Pasadena, suggest that slime
produced by fish and ejected onto the skin surface, reduces friction by
minimising turbulence.  They cite the barracuda as being the most effective
user of this technique, skin friction being reduced by as much as 65.9 per cent.
Similar systems have been tried on marine vessels).

The *Sunday Times* note did stimulate correspondence in that paper's columns,
including a letter from Sir Alliott Verdon-Roe.  Verdon-Roe made the very valid
point that by constructing an ultra-light glider with a low factor of safety[1],
and flying very close to the ground, (A.V. Roe described such a machine as a
'skimmer'), it might have been possible for a man, using arms and legs, to
generate enough power to fly.  He went on to suggest that the introduction of
slow flapping of the wings could further aid performance once take-off had been
achieved.

One letter, written by Frank Lane of Ruislip, pointed out that certain
species of midges sustained a wing beat rate of 1046 per second.  The writer
challenged engineers to reproduce such motion!

In his later paper Shenstone reviewed the work on fixed wing man-powered
aircraft, concluding that 'the state of the art' in 1955 was similar to that
in 1900 regarding powered aircraft.  Flight was possible, but data was lacking
on such features as appropriate wing sections, control and stability, and
power outputs.  By channelling resources into these fields, Shenstone considered
that a successful machine would be forthcoming.

The two different approaches to the question as to whether man could fly
were to some extent incidental.  What was significant was the fact that
Shenstone had been largely responsible for sowing in the minds of the public
a seed of interest in the subject of serious attempts at man-powered flight.

Interest in man-powered flight was also evident at the College of
Aeronautics, Cranfield.  In 1956 a note, described as a preliminary assessment
of the possibilities of man-powered flight, was written by T. Nonweiler.  His
optimistic conclusions, although based largely on slightly misleading figures
for the power outputs of cyclists, were to stimulate academic interest,
initially within his own College.  Nonweiler's research reports and Shenstone's
lectures and articles had a common result - the Man-Powered Aircraft Committee
at Cranfield, and their very significant contribution is described in the next
chapters.

* * *

---

[1] *This refers to the stressing assumptions, and these are mentioned in Chapter 8.
(Correspondence between B.S. Shenstone and W. Tye).*

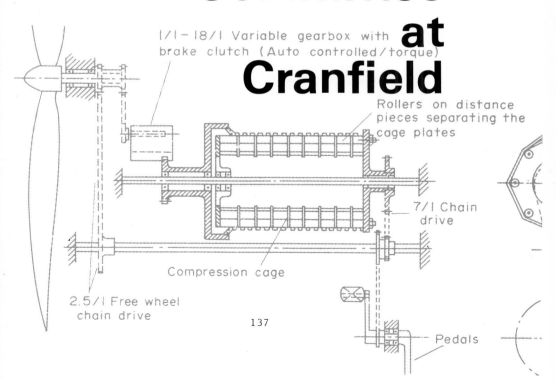

# Formation of the Man-Powered Aircraft Committee at Cranfield

1/1 – 18/1 Variable gearbox with brake clutch (Auto controlled/torque)

Rollers on distance pieces separating the cage plates

7/1 Chain drive

Compression cage

2.5/1 Free wheel chain drive

Pedals

# 8

## THE MAN-POWERED AIRCRAFT COMMITTEE

Much of the credit for the revival of interest in man-powered flight in the late 1950s, particularly in Britain, should be given to the Man-Powered Aircraft Committee (MAPAC) formed at the College of Aeronautics, Cranfield on 10th January, 1957.

The founder members were:-

Mr. H.B. Irving, B.Sc., F.R.Ae.S., President of the LSARA (Chairman)
Dr. J.R. Brown, B.Sc., M.B., B.S., London School of Hygiene and Tropical Medicine
Mr. R. Graham, F.R.Ae.S., M.I.Mech.E., College of Aeronautics.
Mr. Thurstan James, Editor of *The Aeroplane*
Mr. A. Newell, A.F.R.Ae.S., M.I.Mech.E., College of Aeronautics
Mr. B.S. Shenstone, M.A.Sc., F.R.Ae.S., F.C.A.I., A.F.I.A.S., Chief Engineer, B.E.A.
Mr. T. Nonweiler, B.Sc., A.F.I.A.S., Queen's University of Belfast[1].

By February 1958 the membership had been extended to include David Rendel of RAE Farnborough; J.M. Gray, MoS; S. Scott-Hall; R.P. Itter, National Union of Cyclists; Dr D.R. Wilkie, University College, London; Mr. (now Professor) G.M. Lilley and Mr. Jeffrey, both of Cranfield College of Aeronautics; and Peter Scott of the Wildfowl Trust (a keen glider pilot).

The agreed terms of reference, stated in a small pamphlet produced by MAPAC shortly after its formation, were:

(i)   To review published work relevent to the performance of man powered flight;

(ii)  To assess the prospects of its attainment;

(iii) To promote such activities as shall appear likely to lead to its realisation.

As a necessary jumping-off point, the actual definition of a man-powered flight had to be decided, and MAPAC adopted the following form[2].

"Man-powered flight is the act of raising a heavier-than-air craft from the

---

[1] *Now Professor T. Nonweiler*

[2] *The definition of a man-powered flight has been the subject of much debate, and the definition adopted by MAPAC would for example rule out the flights of the Haessler-Villinger aircraft, (see also Chapter 2).*

ground, and sustaining controlled flight, solely by the contemporaneous muscular activity of its occupant(s)."

The definition was to be augmented in due course by the stipulation of a maximum time between start of the take-off run and the achievement of a certain altitude, and the time of flight at this altitude to make it acceptable as a 'sustained flight'.

MAPAC realised that the successful achievement of man-powered flight would be difficult, and proposed international co-operation to widen knowledge, initially in the form of research projects and reviews of past efforts. The Committee acknowledged the fact that the fixed wing aircraft might not be the only effective means for translating power into flight, and also suggested the use of ornithopters, helicopters or autogiros. It also advocated research into power storage devices.

A high frequency of meetings was evident, the fourth gathering being held on 29th October, 1957. Apart from the research work and design studies being carried out at Cranfield, MAPAC was concerned primarily with sounding out opinion on man-powered flight and determining the future path which the Committee should take,

Of interest, and possibly of more current significance[1], was an exchange of letters between S. Scott-Hall, Scientific Adviser to the Air Ministry, and B.S. Shenstone, in May 1957. Scott-Hall wrote to Shenstone, discussing the possible applications of man-powered aircraft: "... Apart from its sporting interest, has the project any possible applications for military purposes? One visualises that the army might be interested from the point of view of mass movement of troops, or dispatch riding." Mr. Shenstone replied in a light-hearted vein: "It has an immediate military value as a more dangerous method of training commandoes". He went on to emphasize the sporting possibilities for man-powered flight, but could have pointed out that Scott-Hall's suggestions might well solve the problem of Service rivalry once and for all!

Shenstone also corresponded with the Royal Aeronautical Society, the first signs of a relationship between MAPAC and this organisation which was to result, within the next two years, in a group being formed within the R.Ae.S. devoted to the topic of man-powered flight. In asking the Society to designate a member of Council to serve on MAPAC, he suggested that such a show of faith would "indicate that we are not necessarily completely crazy, and that our efforts have some sound technical background".

Individual members of MAPAC were responsible for a number of small studies

---

[1] *Three years later man-powered flight was the subject of a comment in the House of Commons by the Parliamentary Secretary to the Ministry of Aviation. Hansard reports him as saying that it was "not without considerable scientific interest" and indicating "the possibility that some new design feature of wide application might arise".*

*Starting in 1971, the United States Air Force have been evaluating very quiet aircraft, to be used on reconnaissance and similar missions. The development of light-weight structures for man-powered aircraft, and the increased knowledge of the aerodynamics of very high aspect ratio wings, both of use here, have been aided by man-powered flight interest.*

on designs for man-powered aircraft.  The Acting Secretary, David Rendel, gave
some thought to the problem of man-powered ornithopters, reporting early in
1958.

He concluded as follows:

"Compared with a fixed wing aircraft, the properties of the ornithopter
have been the subject of a very limited amount of study, in spite of the
fact that over one hundred individual inventors are reported to have built
or planned ornithopters, and half of these appear to have been respectable
scientific endeavours, associated with such famous mames as Handley Page,
Lilienthal, Brustmann, Lippisch, Lindemann and others.  The basic troubles
appear to have been inadequate preliminary thought, unsuitability of
structure, and insufficiently sustained effort and expenditure in the
process of development.  It is clear, however, from (this) analysis, that
the flapping wing does provide a possible means of providing lift at a much
lower power output than any of the more conventional types of aircraft.
For this reason, although it is an unsuitable way of providing lift and
propulsion at high speeds and high powers, it may nevertheless provide an
answer to the problem of a single-man-powered aircraft which the fixed wing
aircraft will not do, vis its dependence on forward speed to obtain its
lift."

Rendel's aircraft was to weigh less than 23 kg and have an all-up weight of
less than 115 kg.  He used these figures in his calculations and determined
that take-off speeds would be of the order of 0.9 - 1.4 m/s and the maximum
cruise speed 3.35 m/s.  (Rendel assumed an instantaneous peak power output of
1.5 kW for a man, using legs, which he considered essential to obtain the
required power).

In March, 1968, MAPAC issued a progress report:

"In the course of the past fourteen months the Committee has met six times.
Members of the Committee are already engaged in specific research projects.
An active experimenter, who has made successful flapping wing models, is
now proceeding with the full scale design.  In this the interest of a
leading British aircraft designer has been enlisted.  Numerous theoretical
studies have been considered by the Committee.

It has been established, in the opinion of MAPAC, that the power is best
provided by some mechanism closely analogous to the bicycle.  It seems
established that the well known pedalling arrangement is the optimum, and
that means of converting this power into thrust is best achieved by the
orthodox airscrew.  MAPAC considers that apart from the individual
experimental work mentioned above, it is now essential that at least one
full scale machine should be built."

In the same report MAPAC put out an appeal for funds to support the
construction of an aircraft, and aimed to obtain £20 000.

## FORMATION OF THE MAN-POWERED AIRCRAFT GROUP
## BY THE ROYAL AERONAUTICAL SOCIETY

On the 28th October, 1958, MAPAC wrote to the Secretary of the Royal
Aeronautical Society, proposing that a Group within the Royal Aeronautical

Society be formed[1]:

"Dear Sir,

In accordance with the Society's memorandum on the formation of Groups, dated Septemper 1958[2], we wish to propose that a Group be formed to study and foster work in the field of man-powered flight.

The aim of the Group is the achievement of flight by human muscular effort alone.  This problem is discussed in a paper by Nonweiler published in the October (1968) number of the *Journal* (official organ of the R.Ae.S.).  The Group intends to achieve their aim by reviewing published work relevant to the subject, by encouraging discussion of ways and means, and by promoting activities likely to lead to its realisation.  The Group intends to organise lectures and discussions, and will form a small committee to do this.  It also hopes to raise funds to enable the work of designing and constructing a man-powered aircraft to be carried out.  For this, it may be necessary to form a small working party within the Group.  However, financial commitments to the Society should be small, and in practice the Group will expect to raise money through its own members for activities in which it is specially interested.

The Group would however hope to receive secretarial assistance from the Society, for example in the production of minutes and notices of meetings, and possibly a short news letter.  It would also hope to use the Society's facilities, such as the library, conference room and lecture room.

> (*Signed* – ten members of MAPAC)."

The above letter, suggesting the formation of a man-powered aircraft group within the Royal Aeronautical Society, was favourably received at 4, Hamilton Place (the headquarters of the R.Ae.S. in London, W.1.), and the inaugural group meeting was held there on October 30th, 1959.  Many of the original MAPAC were present, and the meeting was chaired by Peter G. Masefield, then President of the R.Ae.S.[3]

It was always the intention of the man-powered aircraft group to encourage, with whatever financial support that could be found, the furthering of man-powered flight.  Initially, money was raised by members contributing their broadcasting fees, gifts from companies and small donations from individuals. The financial situation was soon to be greatly eased, however.

---

[1] *It was by no means certain that the affiliation to the R.Ae.S. was inevitable or desired by all the MAPAC members.  The suggestion that a Society for Man-Powered Flight be formed was taken seriously by several MAPAC supporters.*

[2] *A considerable number of these Groups now exist within the R.Ae.S., specialising in many aspects of flight, e.g. Air Law, History, Test Pilots Group.*

[3] *An earlier gathering, on 24th April of the same year, had agreed on the Group formation and formed a Steering Committee, chaired by H.B. Irving.*

## NONWEILER'S POWER MEASUREMENTS

Before moving on from the MAPAC era, reference should be made to some of
the valuable work undertaken, primarily based on Cranfield College of
Aeronautics, by the members of this body[1]. (Following the incorporation of
MAPAC within the Royal Aeronautical Society, work at Cranfield did not stop.
In fact many projects gained momentum. These are the subject of discussion in
Chapter 15.)

Nonweiler's first detailed work was to assess the power that an athelete
could produce, based on estimates of the oxygen intake and the potential energy
available during strenuous exercise. He determined that an exceptionally fit
man weighing 68 kg could produce 0.56 kW over a period of ten minutes, or
0.44 kW over one hour, before complete exhaustion resulted. An average man
would, Nonweiler stated, possibly be able to generate about 0.34 kW over
prolonged periods, but unless he was in training, not much more than 50 per
cent of this power would be produced without undue fatigue. It was also
considered that assistance provided by supplying the pilots with oxygen could
result in a 50 per cent increase in power output.

In his early assessment of the aerodynamic design of a man-powered aircraft,
Nonweiler assumed a span of 18.3 metres, his aim being to minimise power by
aerodynamic design rather than by weight reduction. A wing section with a low
profile drag, the NACA $65_3$-618, was chosen, and using the condition for minimum
power that the induced drag should be three times the profile drag, he arrived
at an optimum aspect ratio of 18 to satisfy this relationship (see also
Chapter 15).

The fuselage drag was determined by assuming that it would resemble a thick
'fin'; this conception was later applied to his full design study of a man-
powered aircraft detailed later in the Chapter.

Nonweiler chose the weight as a variable and determined the power required
to fly using one-and two-man crews for a range of empty weights from 68 to
227 kg. For all weights greater than about 82 kg it proved advantageous to use
a two-man crew, the power per crew member (each weighing 68 kg) being slightly
less than for a single seater. The cruising speed of the two seat aircraft was
of the order of 6.4 km/h faster than that of the single seat machine. He also
used the fact that control and power generation may be an inefficient combination
as an argument for a two man crew.

A direct chain drive from the pedals to ground wheels was recommended, the
latter being connected via a transmission system to the propeller shaft.
Nonweiler considered that a propeller with a diameter of up to 2.75 metres would
be a practical proposition. With a three-bladed propeller of this size, the
rotational speed would be of the order of 250 rev/min.

Before attempting a more detailed design study for a man-powered aircraft,
Nonweiler carried out a number of measurements on the air resistance of racing
cyclists. The results obtained also enabled some interesting data on the power
performance of these cyclists to be calculated.

---

[1] *Some of the studies were started before the formation of MAPAC. Interest at
Cranfield was noted in 1955, and Nonweiler published a preliminary assessment
of man-powered flight possibilities in 1956.*

     A racing bicycle was supported in the working section of a wind tunnel by
wires attached to an overhead balance, as shown in *Fig. 78*. By blowing air at
speeds of up to 17 m/s over the cyclist (who was not pedalling at the time, to
aid stability), it was possible to use the balance to measure the air resistance
as a function of cycling position, speed, and size of rider.

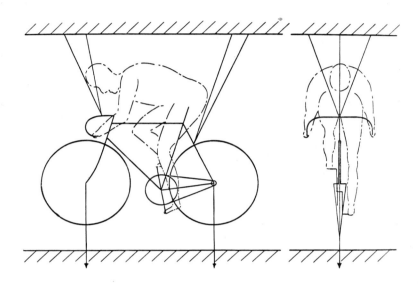

*Fig. 78*  The arrangement used by Nonweiler to determine
air resistance of racing cyclists.  Subjects were suspended
in a wind tunnel.

     Using the drag results for one of the three cyclists tested, a good middle
distance amateur racing cyclist weighing 68 kg, and having a height of 1.79
metres, it was possible for Nonweiler to calculate the powers generated by
pedalling.  Table 6 shows the results of various road race times for the test
subject.  By determining the average speed, knowing the drag from the wind
tunnel measurements, it was simple to determine the power needed to overcome
air resistance.  The power needed to overcome mechanical and rolling resistance,
the friction term in the table, was based on information supplied by the
Raleigh bicycle company.

     These results were encouraging in that they indicated that sustained
powers of almost 0.40 kW could be comparatively easily obtained.  This was at
the time in conflict with tests carried out on ergonometers, where direct power
outputs of trained cyclists were measured.  These latter measurements
suggested that lower powers were achieved, and the diminution of power with
duration of effort, not particularly significant if Nonweiler's data is used,
was quite rapid.  (Nonweiler's results compare quite well with those obtained
by the author's group in 1970 and 1971.  Over durations of ten minutes,
average power outputs by amateur road racing cyclists of between 0.265 and
0.34 kW were recorded).

Table 6.  Power requirements of cyclists

| Distance | | Time | | | Mean Speed | Average Power Needed to Overcome | | Total Resist. |
| km | (miles) | Hrs. | Mins. | Secs. | km/h | Friction kW | Air Drag kW | kW |
|---|---|---|---|---|---|---|---|---|
| 16.05 | (10) | 0 | 23 | 17 | 40.75 | 0.047 | 0.296 | 0.343 |
| 40.13 | (25) | 0 | 59 | 52 | 40.20 | 0.045 | 0.267 | 0.312 |
| 48.15 | (30) | 1 | 14 | 15 | 38.95 | 0.044 | 0.242 | 0.286 |
| 80.3 | (50) | 2 | 7 | 12 | 37.70 | 0.043 | 0.223 | 0.266 |
| 160.5 | (100) | 4 | 30 | 53 | 35.50 | 0.040 | 0.189 | 0.229 |

## THE CRANFIELD DESIGN

The comprehensive design study for a man-powered aircraft completed in 1957
by Nonweiler, with the assistance of J.L. Robinson, at the College of
Aeronautics, Cranfield, was significant because it was the first serious attempt
to utilise most of the disciplines of aeronautical engineering for the design
of an aircraft of this type since the pre-war attempts in Italy and Germany.
A NACA 65A(10)12 wing section was proposed, with an even taper ratio of 5:1
over the semi-span.  With a wing span of 18.2 metres and an area of 15.6 m$^2$,
the aspect ratio was high to minimise the induced drag (which is inversely
proportional to the aspect ratio - see Appendix VI).  In designing the wing
structure a maximum load of three times that of gravity was assumed for
stressing considerations.  Initially, it was hoped to be able to use a foamed
plastic 'fill' with a fibreglass covering for the wing, but it proved to be
easier and more practical at the time to use a rigid birch ply wing skin on a
conventional lattice rib structure, the main strength member being a box
section main spar with two space booms joined by plywood.  The spar weighed
13.2 kg, and the wing covering 11.4 kg, a high figure compared with subsequent
designs with larger wing areas.

A considerable proportion of the design effort was allocated to consideration
of the fuselage layout.  Having decided on a two-man crew to ensure continuity
of power while one member was able to concentrate on control, the tandem
cycling position was chosen to reduce frontal area.  The wheels were encapsulated
within the fuselage, and thus a tall thin structure was required to contain
these features, as illustrated in *Fig. 79*.  In order to keep the resistance to
a minimum, Nonweiler selected a low drag NACA aerofoil section for the fuselage,
as shown in the three-view drawing of his aircraft.  This led logically to an
integral fin, on which was mounted the pusher propeller and tail plane, both
of which were out of the major part of the wing wake.  Although the high
position of the propeller, with respect to the centre of gravity of the
aircraft, would cause stability and control problems, its large diameter
necessary for high efficiency dictated this location.  The propeller diameter
was 2.53 metres, and it had two blades.

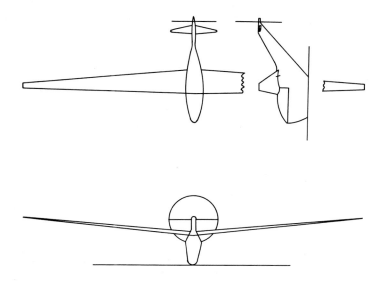

*Fig. 79*  Arrangement drawing of Nonweiler's tandem aircraft

The transmission linkage between the pedals and the propeller shaft had to
extend over a distance of 3.65 metres, but Nonweiler did not enter into great
detail in this field.  Robinson assumed a chain drive from the pedal cranks to
a bevel gear connected to the propeller shaft, but Nonweiler suggested that a
hydraulic system would be appropriate for overcoming the requirement to turn
the drive through ninety degrees.  He appreciated that the efficiency of this
system might not exceed 92 per cent, and any slight weight penalty caused by
the adoption of a chain and bevel system could be more than countered by its
higher 99 per cent efficiency.  An unusual feature of the transmission proposed
was the retention of a three or four speed gear system of the conventional
bicycle type.

Two modes of flight were proposed by Nonweiler, one achieving the greatest
altitude and the second appropriate to the longest flight duration.  The
optimum take-off speed for an altitude attempt of 16.7 m/s, very fast compared
with other man-powered aricraft, would be achieved after a ground run of just
over 910 metres.  At this point the crew would be required to generate 0.95 kW.
The climb following take-off would continue at maximum power, speed gradually
falling off.  Return to cruise power for a short time would complete the climb
phase, and a steady glide to land would follow.  To achieve a flight of long
duration, take-off and flying speed would be that required for minimum power,
and for Nonweiler's proposed aircraft this implied a velocity of 14.6 m/s.  In
order to reduce the induced drag, the flight would be made as close as possible
to the ground.  On the basis of calculations, the maximum altitude of this
aircraft would be 8.2 metres, and the greatest duration of flight about
90 seconds.

In order to increase the power output, Nonweiler recommended that one of
the crew should use hand cranking in addition to pedals.  His work on stability

and control indicated that the present design would require a larger tail area for effective longitudinal stability, and a forward extension of the fuselage for improved lateral stability.

## AN ENERGY STORAGE DEVICE

Many previous man-powered flights had been made with the aid of an elastic launching catapult, or some similar device, and work by W.G. Holloway at Cranfield on a portable energy storage device to be incorporated in the transmission system of the aircraft had the same aim. Holloway's unit, shown in *Fig. 80*, involves the use of rubber stretched in a helix on a cylindrical drum. One end is rotated by the input to stretch the rubber and the other end, when released, provides the output torque. Frictional losses are kept to a minimum by constructing the drum surface as a series of rollers. The storage unit could be used optionally as required, and the torque output could be varied. Input power levels were low, and did not exhaust the pilot before take-off. A typical rubber strand arrangement could provide sufficient power for three minutes flight, but a certain proportion of the stored energy had to be used to overcome the weight penalty of the unit, estimated at 4.5 kg.

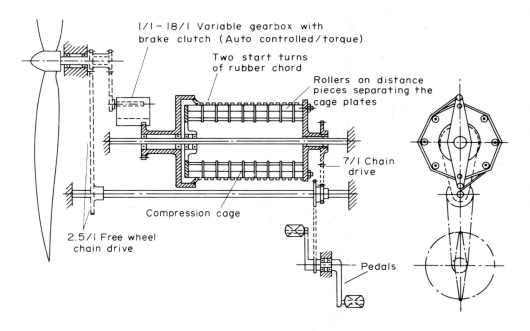

*Fig. 80* A power storage system designed at the College of Aeronautics, Cranfield.

Shortly after completion of his design study, Nonweiler moved to Queen's University, Belfast.

## WILKIE — A BIOLOGIST STUDIES MAN-POWERED FLIGHT

Dr Wilkie, of the Department of Physiology at University College, London, reviewed much of the work of the period on the physiological aspects of man-powered flight, and his conclusions, published in 1959, are of some interest and may be compared with the encouraging results obtained by Günther and Guerra, detailed in Chapter 7.

Wilkie's work in studying the power outputs of animals and birds led to the conclusion that smaller animals have a larger energy output, per kilogram of body weight, than do the larger animals. It was also possible to determine the relationship between energy output and size, and the equation expressing this relationship is discussed later.

Naturally enough, Wilkie stated that the first question to be asked by one studying the ability of man to fly in an ornithopter mode is: "If man can fly, why are there no man-sized birds?" The largest flying birds are the great bustards, weighing up to 14.5 kg (compared with the average weight of a man, which one can assume to be about 70 kg). Before attacking the problem of determining a suitable relationship between the weight and power output of animals, Wilkie produced some interesting results showing the same two variables expressed for aircraft engines, and the results obtained for power output versus weight for piston engines are shown in *Fig. 81*.

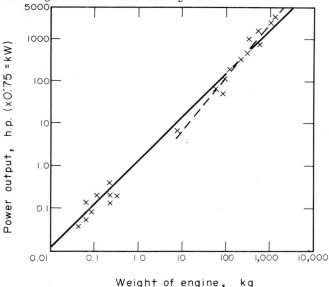

*Fig. 81* Power output of engines as a function of their weight.

The engines shown in the graph all produce about 1 kW/kg. The smallest engines shown are those designed for model aircraft, and it is interesting to note that their power:weight ratio is almost identical to that of the largest aircraft engines. These very small engines are extremely simple, and the maintenance of comparable power outputs per unit weight at the higher ends of

the scale is primarily due to engineering skill.  How successful this is can
be seen in the upper part of the figure, where a dotted line, drawn through
the points representing full sized engines, shows that the power per kilogram
is greater for these examples than for the overall sample, in the former case
the power being proportional to the (weight)$^{1.28}$.

Wilkie presented results for the power output of animals in a similar way,
(Fig. 82).  A variety of features measured for animals are plotted on this
figure, and line A shows the changes in resting metabolism with animal size.
In general, the smaller the animal the greater its resting metabolism per
kilogram.  Measurements made on dogs, horses and man are shown in the figure,
and the resting metabolism may be expressed in units of Watts using an
equation of the form:

Resting metabolism (Watts)  =  3.5 (weight in kilograms)$^{0.734}$.

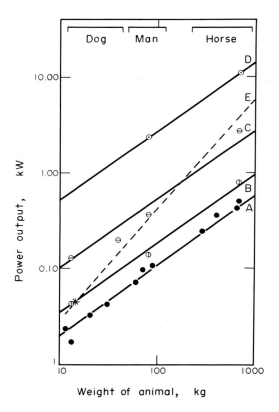

Fig. 82  Power output of animals as a function of their weight.

The resting metabolism expresses energy produced in the form of heat.  A
further set of results is required to express the output of animals when they
are doing external work.  Lines B,C and D in the figure are results of
measurements on dogs climbing a treadmill and pulling a sledge, men cycling and

rowing, and horses pulling.  Each line shows the power output at various levels of effort;  B, maintained all day;  C, maximum steady-state (5 - 30 minutes); D, brief effort of a few seconds.  As the slope of these three lines is identical to that of line A, the above equation, with a different constant, may be used to express the power output of animals doing mechanical work, in terms of their weight.

Results obtained by Nonweiler which showed that the power required to fly for a man-powered aircraft must increase in proportion to the weight in kilograms to the power 1.167 were included as line E.  A condition used in positioning this line was the fact that the great bustard, weighing 14.5 kg, is capable of prolonged flight.  Fitting this bird on line B, with the appropriate weight, enables line E to be drawn.  Any flight requiring power greater than that shown by line C would incur an oxygen debt, and Wilkie concluded from these results that a bird weighing up to 170 kg would be able to fly for a reasonable period. Line E passes just below the maximum steady state point for a man, so he would in theory be able to fly if provided with wings.  However, this does not allow for the fact that any wings and other necessary equipment would be an extra weight, and in accounting for this additional penalty, operation would move above line C, implying a limited duration flight due to oxygen debt.

Wilkie followed Nonweiler in advocating a two-man crew, and proposed a tractor propeller layout in which the pilot contributed some proportion of the total power, concentrating on control, and the second crew member, facing towards the tail of the aircraft, pedalled for all his worth.  Transmission was to be via chains and bevel gear.

## SHENSTONE'S PROPOSED AIRCRAFT

Like Wilkie, B.S. Shenstone also concentrated on one particular aspect of man-powered flight during this period, namely engineering design.  A second common feature was the fact that Shenstone, in 1957, translated his thoughts into a paper proposal for a man-powered aircraft.

His deliberations on the engineering aspects of man-powered aircraft were the subject of the second in a series of lectures given before the Man-Powered Aircraft Group early in 1960[1].  Shenstone considered the paramount engineering problems to be:

  (i)   the arrangement of the structure,

 (ii)   the details of the structure,

(iii)   power transmission.

All of these would of course be affected to a greater or lesser extent by aerodynamic requirements.  These apply particularly to the wing, but detail design of the fuselage and propeller location could have very significant consequences on the drag, and hence the power required to fly.

A fixed wing propeller-driven layout was adopted by Shenstone from the

[1] *B.S. Shenstone.  'Engineering aspects in Man-Powered Flight'.  J.R.Ae.S. Vol. 64, pp 471 - 477, August 1960.*

start because immediate knowledge in depth was available on the performance and
aerodynamics of such a combination applied to powered aircraft, and indeed to
man-powered machines.  Such was not the case with ornithopters.

In common with most other designers, Shenstone considered that a high
aspect ratio wing would be used, but thought that this should offer no particula
structural difficulties.  The choice of a high or low wing location with respect
to the fuselage was open to debate.  Those proposing high wings argued that the
disadvantage of having a wing slightly further away from the ground, thus
possibly not coming within the beneficial 'ground effect' region where drag is
reduced, would be outweighed by the facts that the wing would be less liable
to damage, and it would be easier to ensure satisfactory aerodynamic characteristics near the wing/fuselage joint.

Shenstone felt that it was necessary to consider the propeller and tail
unit as a single entity.  (While this may be ideal, it is often not achievable.
It is however, important to ensure that the flow of air through the propeller
does not affect the flow over the wing.  In other words, the slipstream must
be remote from the wing.)  The most common ways for achieving this have been to
mount the propeller on a pylon above the wing, or to site it at the tail.  The
Haessler-Villinger aircraft was an example of the former, while the Hatfield
'Puffin' of the 1960's adopted the second layout.  Shenstone also suggests the
use of twin tail booms with the propeller centrally located, or the design of
an annular type unit which would externally resemble part of the fuselage.
(High efficiencies needed dictate that the propeller must be generally at least
1.8 metres in diameter).

The simplest structural materials, spruce and balsa wood, would, Shenstone
believed, prove adequate for serious designers of the late 1950s.  He did,
however feel that the inclusion of some form of honeycomb or plastic foam
structure may be beneficial in providing structural rigidity and maintaining
the aerofoil profile so necessary for good aerodynamic efficiency.  A very
light covering for the wing would be required, consistent with the preservation
of the wing form.

## THE STRUCTURAL STRENGTH OF MAN-POWERED AIRCRAFT

In this context, it is worth noting that Shenstone had corresponded with
Walter Tye, then Chief Technical Officer of the Aircraft Registration Board
(ARB)[1], a body responsible for the stipulation of overall performance criteria
for civil aircraft.  In April 1958 Tye produced a note entitled "Stressing
Assumptions for Man-Powered Aircraft", in which the ability of such machines to
cope with gusts, dives and other events, even when flying a few metres from the
ground was examined.

Tye considered that over flat ground at low altitudes vertical gusts could
be neglected, but he believed it would be wise to take into account horizontal
gusts of up to 16 km/h.  The diving speed would be restricted by the fact that
the aircraft might only be flying at an altitude of five metres, and Tye felt
that a speed build-up in excess of 24 km/h in a dive so close to the ground
would be unlikely to occur.  Normal accelerations during manoeuvres by the
pilot might impose additional loads on the structure, (a jet fighter, for

---

[1] *Now the Civil Aviation Authority*

example, must be able to withstand normal accelerations amounting to several times that of gravitational acceleration). These could be covered by an increment of 0.5 g, giving a maximum acceleration of 1.5 g, where g denotes the acceleration due to gravity.

The normal practice with powered aircraft is to apply a factor of safety on the ultimate loading due to accelerations. This factor is conventionally taken as 1.5 times the limit load. This does not cover inaccurate stressing, which can be revealed on powered aircraft by structural tests. Tye considered it unlikely that sufficiently rigorous structural experiments would be carried out on man-powered aircraft specimens, and suggested an extra ten per cent safety factor, giving an ultimate load factor of 2.5[1].

While retaining sufficient strength to meet the requirements of safety, Shenstone felt that the fuselage, apart from the load-bearing points carrying pilot, transmission and the wings, should be a very light streamlined fairing.

Commenting on the transmission systems, Shenstone made the point that, except during sprint conditions, (durations of less than two minutes), there is no advantage in using more than one's legs. This is based on the argument that the full cruising power of a person is limited by his ability to use oxygen, and this limit is reached before that due to the leg musculature. Pedals, chains and belts were noted as the preferred methods of power transmission, but Shenstone appreciated that the system adopted was influenced by aerodynamic considerations to a very great extent.

In concluding his summary of the pertinent engineering features of these machines, Shenstone makes the very valid point that success will only come as a result of design, manufacture and testing *par excellence*, rather than from some flash of inspiration. On an even more realistic note, he states: "We have not even the consolation that our powerplant will improve in (the) foreseeable future".

Shenstone's proposed man-powered aircraft, conceived five years before his paper on engineering aspects of design, had a two-man crew, and was unique in that the propeller was mounted on a driving ring which moulded into the fuselage skin. Transmission to this ring from a chain drive was via a miniature gearbox. He visualised a tow or catapult launch for his aircraft, which was to have an all-up weight of 221 kg and a wing span of 18.3 metres. Wing-warping was advocated in lieu of control surfaces, and, surprisingly, boundary layer control using perforations in the wing upper surface was also incorporated. Provision for sucking away the boundary layer was *à la Raspet*, the suction being created by propeller rotation.

*  *  *

MAPAC, in their efforts to encourage world-wide collaboration and exchange of information on man-powered flight, wrote to the Soviet Union, via the Russian Air Attaché in London. Information was requested concerning a reported man-powered aircraft which had flown in Russia[2], but it was a year later before a reply was received. Even then no mention was made of the aircraft, but the

---

[1] *The ultimate load factor is obtained by multiplying the limit load factor by the safety factor and the stressing inaccuracy factor (i.e. 1.5 × 1.5 × 1.1 = 2.5).*

reply did indicate that man-powered flight was being studied, two configurations being shown in *Figs. 83 and 84*. The letter originated in the Aeronautical Research Centre, Moscow, and was signed by A. Zhukov, President of the Committee on Man-Powered Flight, and E. Vinogradev, Secretary.

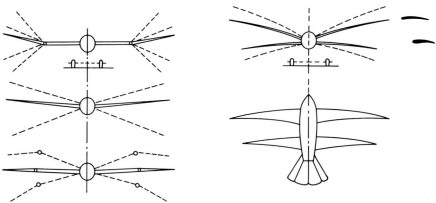

*Fig. 83*  Various Russian ideas on ornithopter layouts I.N. Vinogradov.

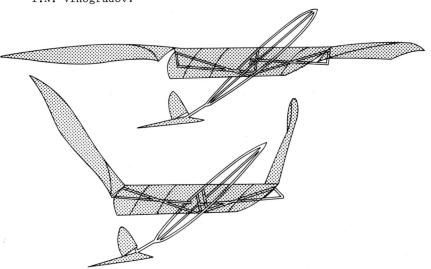

*Fig. 84*  V. Yakovlev's flying model of an ornithopter, 1948. Weight:  10 g;   Span:  810 mm.

---

2 *In the 'Vancouver Sun' of December 15th, 1958, the following note appeared: "Russian inventor Anatoly Invanyuta has developed a bicycle which can fly at over 40 m.p.h. (64 km/h), the Soviet News Agency, Tass, reported today".*

*Shenstone, in a letter to a correspondent who had sent him this news clipping, indicated that similar reports often appeared in Russian newspapers and other sources, but all claims were so far without foundation.*

*9*

UNIVERSITY SOUTHAMPTON
FIRST MAN-POWERED FLIGHT GT BRITAIN
9TH NOVEMBER 1961

# The Kremer Competition

*– Catalyst for Worldwide Activity,*
*and the First Entrant*

*9*

## HENRY KREMER — THE KREMER COMPETITION

A letter which was to have a major influence on the development of man-powered aircraft, particularly in Great Britain, was published in the January 1960 issue of the *Journal of the Royal Aeronautical Society*. This letter, written by Robert Graham, a leading figure in the activities of the Man-Powered Aircraft Committee, later to become the basis of the Man-Powered Aircraft Group of the Royal Aeronautical Society, contained the statement that Henry Kremer, a London company director, had offered a substantial prize for the first successful flight of a British man-powered aircraft, and read:

"I am very pleased to be able to inform you that Mr. Henry Kremer of 9 Kingsway, London W.C.2, Chairman and Managing Director of Microcell Limited and also a Director of B.T.R. Industries and other Companies, has most generously offered to give a prize of £5000 (Five Thousand Pounds)[1] for the first successful flight of a British designed, built, and flown Man-Powered Aircraft, such flight to take place within the British Commonwealth, under conditions laid down by the Royal Aeronautical Society.

Mr. Kremer has authorised me to write this letter and to say that information regarding this prize, and the conditions governing its award, may be made public at your convenience.

(Signed)   R. Graham."

This offer, first announced by the Royal Aeronautical Society in November 1959, was to give a further stimulus to the growing desire to solve the problems of man-powered flight. Under the guidance of the Man-Powered Aircraft Group of the Society, in conjunction with the Royal Aero Club, a set of rules were formulated for what had become known as the Kremer Competition, and these are given in their latest form, as published in January, 1976, in Appendix V.

The direct implications of the rules will be considered later.

At a Society meeting, Mr. P.G. Masefield, acting as President of the Royal Aeronautical Society, in accepting the prize, thanked Mr. Kremer for his "very generous offer" to donate a £5000 prize. He described the offer as "most exciting news and a great incentive to the Man-Powered Aircraft Group". He also said that a meeting of the Royal Aeronautical Society was to be held in the immediate future to consider the conditions to be laid down for the prize.

Further financial incentives were received by the Royal Aeronautical Society. Air Commodore J.G. Weir gave a cheque for £1000, stating

---

[1] *The Kremer Competition prize has since been raised to £50 000 – See Appendix V*

that the money could be used by the Group Committee in any way that would be of
benefit to the development of man-powered flight.  This sum was to form the
basis of a fund to assist the development and construction of promising designs
submitted to the Society.  When Mr. Henry Kremer was informed of this fund he
made a further considerable contribution to it, as did the three leading
aircraft companies in the United Kingdom.  Thus by the middle of 1960 almost
£5000 was available for financing both basic research and the construction of
actual aircraft, and by March 1961, the following major contributions had been
made to the Man-Powered Aircraft Group fund (excluding the prize money):

| | |
|---|---|
| Mr. Henry Kremer | £2500 |
| Air Cdr. J.G. Weir | £1000 |
| British Aircraft Corp'n. | £ 500 |
| Hawker Siddeley | £ 500 |
| Shell & B.P. | £ 100 |
| Westland | £  50 |
| B.P. Trading | £  31.50 |

In addition, members of the Group donated proceeds obtained from fees for
papers written by them, and a number of personal contributions were made.

The Kremer Competition prize was offered for the first man-powered aircraft
to complete a figure-of-eight course around two markers one half of a mile
(803 metres) apart.  The rules governing the flight conditions may be stated
briefly as follows:

The power for take-off and during flight has to be solely muscle power,
and the aircraft must cross the starting and finishing lines at an altitude of
not less than ten feet (3.05 metres).

The rules embraced Commonwealth entrants, and considerable contributions
to the science of man-powered flight were later to be made by Canadian
aeronautical engineers.

Reference to Appendix V , which contains the latest rules in full, shows
that the regulations are directed at limiting the award of the prize to design
teams who are capable of producing and flying an aircraft which could meet what
appears to be a very severe interpretation of the term 'man-powered flight'.
In fact, after the first Annual General Meeting of the Royal Aeronautical
Society Man-Powered Aircraft Group, a meeting of the Group Committee was told
by Mr. Graham that in discussions with Henry Kremer, the latter had said that
several people had commented that the conditions for the award of the Kremer
Prize were too arduous.  Kremer had suggested that if a controlled flight was
made, but was found not to quite conform to the rules of the competition, it
should be within the power of the Committee to consider the flight on its own
merit, and possibly make the award.  He also considered that the Committee
should be able to change the rules governing the flight if they proved in
practice to be too severe[1].

---

[1] *A.M. Lippish, commenting on the Kremer Competition rules, states:  "As in
anything ... new, the start and the initial effort are painstaking.  It is
therefore wrong if we put up a goal which cannot be reached by such initial
effort.  This happened with (the German Prize in 1936) and it seems to me
that the Kremer Prize of the R.Ae.S. is in the same category.
If anyone thinks about ... an American man-powered (flight) competition, I
hope that we will be less ambitious ..."*

Mr. Graham had in mind the case of a competitor who did not turn at the correct point, this being a very slight deviation from the rules. It is likely that Mr. Kremer was thinking of much more radical variations in the interpretation of the course. The meeting at which these questions arose was held on 5th May, 1961, several months before any of the potential Kremer Competition entries took to the air.

Provision was made for the review and amendment, if necessary, of the conditions governing the award of the prize, and this will be commented upon later.

## IMPLICATIONS OF THE RULES

It is worth examining the condition(s) of entry, (Section 4), in turn. Their main implications will be explained, and the final section of this chapter will describe the efforts of the design team at Southampton University in translating their interpretation of the regulations into hardware.

The first sub-sections deal with the aircraft, and the interpretation given here is based, where necessary, on the assumption that a conventional fixed wing aircraft, as opposed to an ornithopter or helicopter, is to be used in the attempt. Earlier discussions on the relative merits of these three craft suggested that the fixed wing aircraft was the most suitable configuration at present for achieving sustained man-powered flight.

"... the machine shall be a heavier-than-air machine ... use of lighter-than-air gases shall be prohibited".

The rules prohibiting the use of man-powered balloons, or the filling of wings with bags of helium or some similar low density gas, directed the structural designers to search for solutions to the problem of providing ultra-light structural members with substantial strength, with strength to weight ratios well in excess of, for instance, those of typical glider components. Initially this necessitated that emphasis was placed on obtaining a structure formed mainly of spruce and balsa wood, designed for optimum strength whilst maintaining very low weight. This form of construction was applied to all the main parts of the aircraft; wings, fuselage, tail surfaces etc. Very light fabric or plastic materials were chosen for covering these members, and with them the Haessler-Villinger and Bossi-Bonomi aircraft achieved significant weight improvements over contemporary gliders in the 1930s. Recently the availability of man-made rigid foam materials and the filament-wound fibre-resin composite materials has widened the designer's scope considerably.

Designs using combinations of the light woods, low density metal alloys, and a variety of skin materials have proved to be an order of magnitude lighter than gliders of comparable overall size, and these achievements are compared with the weights of comparable single-seat gliders in the graph in *Fig. 85*. (It must be remembered that gliders have to endure flying conditions much more arduous than the average man-powered aircraft).

"The machine shall be powered and controlled by the crew of the machine, over the entire flight ... no devices for storing energy either for take-off or for use in flight shall be permitted".

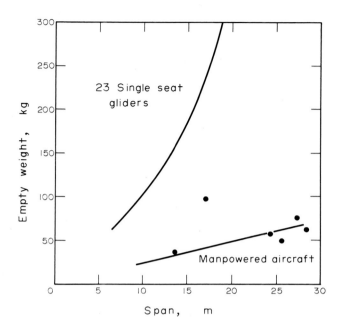

*Fig. 85* A comparison of the structural weights of some
current gliders and man-powered aircraft.


In Chapter 8, reference was made to the use of energy storage devices in
man-powered aircraft, particularly the system outlined by T.F. Nonweiler.
These could be tensioned before take-off, possibly by the pilot himself, and
would provide useful energy at the expense of a small increase in all-up weight.
No aircraft appears to have flown incorporating its own energy storage device,
although catapults have been used to assist launching.  Thus their effectiveness
is difficult to assess, but some form of assistance during take-off, when the
pilot may be tempted to expend a considerable proportion of his energy, would
probably prove a valuable boost to performance.

Prohibition of energy storage devices simplified the problems of
designing man-powered aircraft to some extent in that one important variable
was to all intents and purposes eliminated – this being the power available to
take-off and complete the course.  The power produced by a man, both as a
professional athlete and as an 'average fit' person, in various forms of
locomotion, has been fully covered in Chapters 5 and 8.  It has been shown that
designers of aircraft for the Kremer Competition had a wealth of information to
hand on this topic.  As a result they were able to concentrate on translating
this power most efficiently into a thrust, and also on determining the optimum
number of crew required to do this, solutions to the latter problem varying
between a single man and a 'rowing eight'.

Perhaps the most significant part of the conditions covering the crew duties
was the requirement that they control the aircraft.  In later descriptions of
projected aircraft, it will become evident that design calculations as to the

durations of flight and the abilities of the crew to complete the Kremer course appear to have been optimistic. Early potential Kremer Competition entrants may not have fully appreciated the implications of control, and although the topic is discussed elsewhere, mention should be made of the fact that the majority of the more accessible results on the power outputs of man in, for example, the cycling position, were obtained under ideal conditions. 'Ideal' in this context refers to the concentration available for pedalling. The control of a bicycle is, with practice, largely a reflex action, and the rider is free, on the race track or in a road race, to devote most of his attention on realising maximum speed and optimum use of his muscular energy. A pilot of a man-powered aircraft, however, cannot in general be expected to exhibit similar reflex control type actions in his flying of the aircraft. This is particularly so when it is remembered that many man-powered aircraft, although inherently stable in roll, show pitch instability which has to be countered continuously. The pilot is not aided in this by his frequent lack of external reference points, which would give him quick indication of the attitude of the aircraft[1].

"No part of the machine shall be jettisoned during any part of the flight including take-off."

This condition is primarily aimed at ensuring that the undercarriage is retained throughout the flight. Initial acceleration is, as a rule, generated by direct drive from pedals to a wheel in contact with the ground, the propeller contributing little thrust at this stage. The propeller is soon able to take over the function of maintaining sufficient thrust to enable the aircraft to become airborne, and sustain it in this state.

Although a small amount of weight could be saved by release of part of the undercarriage once free flight has been achieved, it has become evident that flights of man-powered aircraft are often interrupted by returns to the runway for short periods at the initial flight stage, and the ground wheel may then be required to gain further momentum for a sustained flight. An undercarriage of substantial construction also protects the aircraft from damage.

The author's design team proposed to incorporate outer wing sections which would break off if they came into contact with the ground or a pylon, this being intended to minimise the extent of damage to the rest of the airframe. Crashes in which this feature could prove effective are wing-tip stalls and collisions. Should any need to use these devices arise, the Competition attempt would in all probability be abortive, and their inclusion serves only as a safety factor.

Were components allowed to be jettisoned, one possible gain could accrue to the ability to carry ballast. It is usual practice to hold an aircraft on the ground after sufficient speed has been achieved to enable lift-off to occur. This assists in obtaining a good rate of climb, and is implemented by the control surfaces. The use of ballast in a man-powered aircraft would help to hold the craft on the ground until a higher speed than is considered normal for take-off was obtained. Sudden release of the ballast would then result in a high rate of climb, at least initially, and the additional power used during

---

[1] *Experiences of the pilot of 'Puffin', the Hatfield aircraft recounted in Chapter 10, do not fully support this view.*

the take-off run would probably be more than compensated for by the extra
metres gained in the levelling-off altitude. The low glide angle of these
aircraft would ensure that any advantage thus accomplished was not immediately
lost.

Regulations governing the crew are straightforward, the permission to use
a ground handler during take-off being necessary in view of the large wing
spans involved and the fact that the wings are likely to have pronounced anhedral
before sufficient lift is obtained to balance their own weight.

Ground conditions specified for the Kremer Competition, although eliminating
the obvious advantages to be gained by taking-off down hill into the wind, were
also designed to protect man and machine. The lightly stressed structures
designed for minimum weight are unable to withstand strong gusts of wind, or
any sudden accelerating force, and serious accidents could occur if flights
were attempted in adverse weather conditions. The early morning is a favourable
time of day for flying, although one tends to lose the advantage of warm
convective currents rising from a sun-scorched runway. Pilots of medium-sized
piston engined aircraft have often reported difficulties in setting the
aircraft on the ground at normal landing speeds in hot climates when the after-
noon sun has produced just this effect.

## HOW TO FLY THE 'FIGURE OF EIGHT' COURSE

The 'figure of eight' course proposed in the original Kremer Competition
has to date proved impossible to fly, and in fact no official attempts were
made up to the beginning of 1970. It has been realised that the design of the
aircraft must be influenced by the shape of the course to be flown, and
numerous variations have been put forward as an 'idealised' flight path.

For a small aircraft with an engine power of, typically, 150 hp (114 kW)
large angles of bank could be tolerated before losses in altitude during this
turning manoeuvre became dangerous, and the form in which it might attack the
Kremer course is shown in *Fig. 86*.

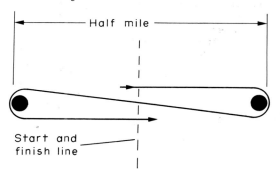

*Fig. 86*  The Kremer course as it would be attacked by
a small powered aircraft.

The flight necessary would be little over one mile (1610 metres) in length,
plus the take-off and landing runs.

For the man-powered aircraft, however, restrictions on the angle at which turns can be made, brought about by the large wing spans and limited power, and the low altitudes achieved mean that the flight path must follow forms similar to those shown in *Fig. 87*.

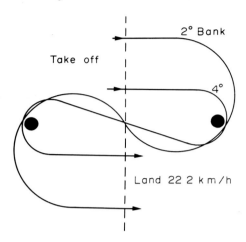

*Fig. 87*  Lorne Welch's proposals for attacking the Kremer course.

Lorne Welch, using data for an aircraft with a wing span of 24.4 metres, has suggested[1] that a course with long straight sections should be followed using an angle of bank of four degrees in the turn.  Considerable stretches of straight and level flight would be possible, and the avoidance of steep turns would ensure that the aircraft could be flown close to the ground, making full use of ground effect.  A turn with a two degree angle of bank would involve straight runs at the commencement and termination of the course, but accurate control would be required during the major part of the flight.  It must be remembered that turning flight is in practice by no means the most effective way for generating lift, and power requirements during this phase of the flight are likely to be substantially increased.  Pilot concentration will inevitably be greater, and the danger of tip stalls is always present.

A simple mathematical relationship exists for determining the radius of turn for small angles of bank:

$$R = V^2/g.\tan \phi$$

where  V  is the aircraft velocity,
       $\phi$  is the bank angle, and
       g  is the acceleration due to gravity.

Using this formula to obtain  R,  the radius of turn, for various angles of bank and aircraft speed, the graph in *Fig. 88* is obtained.

[1] *Lorne Welch.  'Gliding and Man-Powered Flight'.  J.R.Ae.S.  Vol. 65, pp 807 - 814, December 1961.*

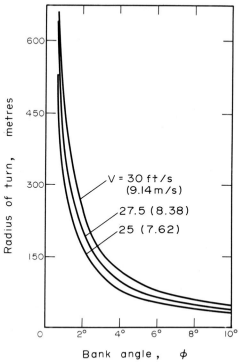

Fig. 88  The angle of bank of a man-powered aircraft,
and the corresponding radii of turn.  Spillman.


It should be noted that the radius of turn decreases very rapidly with
increase in bank angles when the angle of bank is very low.

Although this would suggest that a shorter course could be flown if values
of φ of about six degrees were attained, Spillman pointed out another major
drawback in attempting this method of attack.  He determined that in order to
counter the larger induced rolling moment caused by an uneven distribution of
lift over the wing when turning, aileron powers greater than those on a Viscount
airliner would be required to keep the bank angle steady.  The extra drag
associated with this would be intolerable.

It was suggested that a man-powered aircraft with a span of 27.5 metres
would require a maximum bank angle of 1.4 degrees, if these excessive control
contributions to the drag were to be kept within acceptable limits.  Spillman's
course, shown in the second of the two drawings, Fig. 89, has a length of
3.6 kilometres, and no steady level flight would be possible unless the course
length was increased above this minimum practicable figure.  (Since the Kremer
Competition was first announced, an additional 'slalem' course has been
included, and this is described in Chapter 10 and Appendix V ).

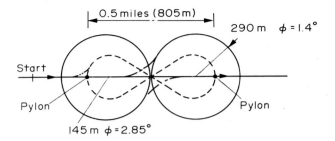

*Fig. 89*  Spillman's minimum practicable 'figure of eight'
for an angle of bank of 1.4 degrees.

## IS IT TOO DIFFICULT?

In general the regulations governing the award of the £5000 prize have
proved severe, and do appear to have been directed at the ultimate in man-
powered aircraft design, with very high propulsive and aerodynamic efficiencies.
With hindsight, the completion of the new 'slalem' course would have been an
ample initial aim, and several authors of papers on man-powered flight in the
technical journals have consistently stressed the amount of work needed, much
of which could not be performed by a single team, before man-powered flight
could become a reality in the context of the Kremer Competition rules.

The controversy surrounding the rules finally adopted for the Kremer
Competition was considerable.  It is interesting to note that in a memorandum
dated 13th November, 1959, Shenstone suggested a set of conditions for the man-
powered flight prize proposed by the Man-Powered Aircraft Group.  They were as
follows:

1.  The flight should be in a straight line.
(Turns are likely to require more height than could be attained with
early aircraft).

2.  The distance should be one mile (1.61 km).
(Here Shenstone believed that a lower distance would not be a
sufficiently high goal, and the Haessler-Villinger machine should be
bettered.  He estimated a flying time of 2.5 minutes, enabling a
continuous power rating in excess of 0.5 horsepower (0.38 kW) to be
maintained.  Also, the distance was well within the maximum length
of existing runways.)

3.  The flight should be over level ground.
(Shenstone put forward the suggestion that 'level' might be interpreted
as not varying more than ten feet (3.05 metres) over the course).

4.  Take-off must be defined to cover the various modes, and flight
length varied accordingly.

   4.1  Take-off by direct crew power - flight of three-quarters
   of a mile (1.2 km) from unstick.

4.2  Take-off by stored energy, generated by crew not more than thirty minutes prior to flight - one mile (1.61 km) from unstick.

4.3  Take-off by external source (e.g. catapult) - 1.25 miles (2 km) from unhooking.  (This was governed by the speed generated by the external source, and the altitude given to the aircraft).

5.  The aircraft would have to fly level for some distance before crossing the starting line.
(To obviate pilots taking advantage of a dive - only applicable to 4.2).

6.  The flight must be observed by Officials of the Royal Aero Club[1].

Villinger, writing in *The Aeroplane* on 6th April 1961, also suggested that it would be preferable to offer a prize for a flight of five hundred metres in a straight line - the Kremer Competition was too difficult with the present state of the art.

It appeared that several members of the original MAPAC thought that the regulations adopted for the Kremer Competition were far too strenuous, and that it would be several years before a flight conforming to these rules could be achieved.

Doubt was expressed regarding the Kremer Competition rules by Thurston James, then Editor of *The Aeroplane* and a founder member of MAPAC, in the edition of February 5th, 1960.

"Has the Royal Aeronautical Society, with the help of the Royal Aero Club, put the magnificent prize of £5000 out of reach of enthusiasts who are striving to achieve man-powered flight?

... As the latest thinking indicates that the accelerating thrust will have to be obtained from wheels rather than an airscrew, it seems a pity that the Royal Aeronautical Society has forbidden the use of a launching trolley. This is something that the motorless flight experts have been using for years.

We wonder, too, how much the necessary insurance premiums are going to cost. It is hard to imagine much third-party damage being done on a windless aerodrome by a machine weighing a few hundred pounds.

In our view the rules are calculated to produce a really worthwhile machine. But it may take years to develop it.

Has the prize been put too far out of reach?"

*   *   *

---

[1] *Now changed - see Appendix V.*

## THE SOUTHAMPTON UNIVERSITY MAN-POWERED AIRCRAFT

At 4.00 p.m. on the 9th November, 1961 the Southampton University man-powered aircraft took to the air after an unassisted take-off *(Fig. 90)*, and flew a distance of sixty-four metres, being airborne for approximately eight seconds.

This was the first wholly successful unaided man-powered flight to take place in Britain, no catapult or tow being used for the launch. Take-off occurred at a speed of 32 km/h after a gradually accelerating ground run.

Further details of this flight are well worth recording. The pilot on this occasion was Derek Piggott, the Chief Flying Instructor at the Lasham Gliding Centre, where the flight took place. No wind was blowing and the day was dry, optimum conditions for testing such an aircraft. The Southampton University man-powered aircraft (SUMPAC), having completed its take-off run, climbed quickly to a height of about 1.8 metres and flew in a stalled attitude at this altitude over a distance of 15.2 metres. A rapid descent followed, the starbord wing hitting the ground first, resulting in a minor ground loop.

Preliminary discussions at Southampton University on the subject of designing a man-powered aircraft were held in the Spring of 1960, and design work on a major scale commenced in July of that year. The principal designers of the aircraft, Anne Marsden, David Williams and Alan Lassiere, supported by the Department of Aeronautics, (now the Department of Aeronautics and Astronautics) under Professor G.M. Lilley of the University, completed the project study at the end of September 1960, and a report was submitted to the Man-Powered Aircraft Group Committee of the Royal Aeronautical Society in the hope of receiving financial assistance from the fund set up to support such projects. Construction commenced in January 1961, and a promise of a grant from the Royal Aeronautical Society was received the following month. The grant amounted to £1500.

It is important to note here that the Man-Powered Aircraft Group Committee comprehensively reviewed progress on man-powered aircraft, and the work of the various teams were closely scrutinized before grants were given.

Some of the comments of the Committee concerning the Southampton proposal are worth inclusion.

Rendel, writing in November, 1960, recommended wind tunnel tests to check the suitability of the wing covering on the machine. He suggested that it be pointed out to the group that it was not the roughness or waviness of the wing surface that would create problems at the low Reynold's numbers (here implying low speeds) concerned, but the ability of the covering to maintain an accurate wing profile. He also stated that the general design and layout were of merit and suggested encouragement for the Southampton group.

In January 1961 a full committee meeting approved a grant of £100 to the Southampton group to finance wind tunnel tests, but before giving further support they requested information on the drag and wing performance.

The designers worked to a specification which would meet the requirements of the Kremer Competition, this giving a useful guide when fixing the main parameters.

Included in the preliminary information required to enable the aircraft size to be determined is the power available. Having rejected configurations such as the helicopter and ornithopter, these types of craft having efficiency disadvantages, a high wing monoplane of large aspect ratio was chosen, a three-view drawing of which is shown in *Fig. 91*. The power available clearly depends upon the number of crew and their method for translating muscle power into thrust through a propeller. For ease of control and drag advantages, in that

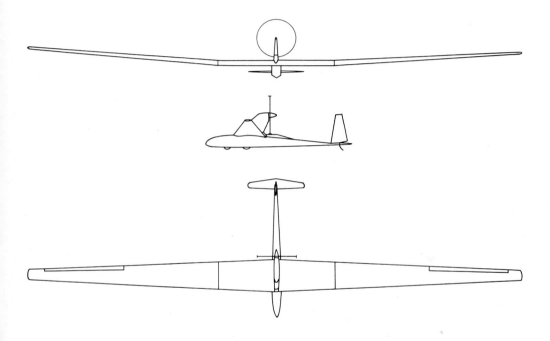

*Fig. 91*  Three-view drawing of SUMPAC

the fuselage frontal area can be considerably reduced, a semi-reclining pilot position was chosen, and this is clearly visible in the seat arrangement *(Fig. 92)*.

Southampton based the preliminary calculations on power availabilities of 0.42 kW for the first minute, i.e. during the take-off run, 0.34 kW for a further four minutes, and an approximately constant power output of 0.305 kW at 80 per cent efficiency for the remaining flying period. They determined that the duration of flight could be up to one hour. Using a simple equation describing the power required for level flight at sea level, the relative importance of the main design parameters may be inspected. Denoting the power by P, the equation approximates to:

$$P = \frac{3.0}{\eta_o} \, W \left(\frac{W}{S}\right)^{\frac{1}{2}} A^{-\frac{3}{4}} \, C_{D_o}^{\frac{1}{4}}$$

where   $n_o$   is the total propulsive efficiency,
        W    is the all-up weight (kg)
        S    is the wing area (m$^2$)
        A    is the aspect ratio
        $C_{D_o}$ is the profile drag coefficient.

This equation shows that the requirements for low power are, in descending order of importance, high efficiency, low weight, high aspect ratio, low wing loading and low profile drag. By manipulation of this equation, an expression for the power required per crew member can be obtained. The Southampton team concluded from the results that one crew member was the optimum if the minimum power criterion was applied. Other advantages cited were slower speed, hence smaller turning circle, and the fact that the crew would be more readily

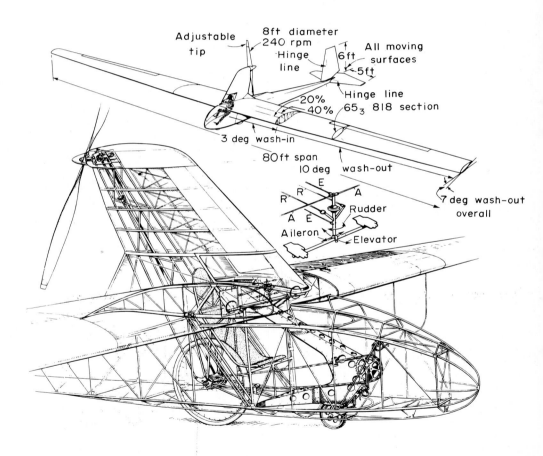

*Fig. 92* The Southampton University aircraft, showing cockpit and transmission details.

available, *(Fig. 93)*. A span of 24.4 metres, with a wing area of 27.8 m² was chosen. The span was limited because it was thought that difficulty would be encountered in turning the aircraft close to the ground. Both the Southampton and Hatfield groups later agreed that spans of the order of 30 metres should be manageable, and 'Puffin II' flew with a wing span of 28.35 metres.

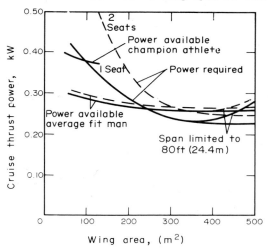

*Fig. 93* SUMPAC – The power available and required for flight (per crew member)

Further features of the wing have an influence on the power requirements. The thickness to chord ratio partially determines the wing strength, weight and profile drag, and a compromise has to be reached between the conflicting requirements of high strength and low profile drag. The Southampton aircraft initially used a single laminar flow wing section with a relatively high thickness to chord ratio for the complete wing span, visible in *Fig. 94*.

The taper ratio used on the Southampton aircraft was 4:1. The choice of aerofoil section for the wing was directed at minimising the possibility of wing root, or, more important, wing tip, stall. Although as mentioned one wing section was originally proposed for the whole wing, the group later adopted three different sections between the wing tip and root. The associated stalling characteristics were thence such that the complete wing tended to lose lift at the same incidence, eliminating tip stalls as such, at least in the straight and level flight condition.

The fuselage of the Southampton machine had three primary functions:  a streamlined structure to contain the crew;  a stable platform for the transmission system;  support for the rudder, tailplane, and other control surfaces and their associated levers.

Drag calculations were made for various values of fuselage length to diameter ratio, and an optimum size selected consistent with the requirement for low drag and the length of fuselage needed to provide effective rudder moment arm for control:  a fuselage of 7.61 metres length, with a maximum cross-sectional area of 0.475 m², was adopted. Tailplane and fin dimensions were

dictated by surface areas required for stability and control, and these
components were all-moving.  The tailplane span was 3.05 metres and the fin
had a height of 1.83 metres.  Their effectiveness was such that movements of
five degrees were sufficient to ensure complete control.

Considerable problems were encountered in positioning the propeller pylon
above the wing, where interference between these two members caused flow
separation and the inevitable high drag.  As the construction programme was too
far advanced at this stage to alter the pylon position, various fairings were
tried in an attempt to reduce the interference drag.  An additional fairing
which reduced unfavourable characteristics to negligible proportions was
evolved.

The structural design of the primary members, the wing, fuselage, tail
unit and pylon, was governed by the minimum weight requirement.  At the time
of the construction the impact of new materials such as foam plastics and the
like had not reached today's proportions, and carbon fibre technology was not
available.  As construction techniques using light woods such as spruce and
balsa were proven, these were adopted for a large proportion of the structure.

Structural design of the wing was aided using a typical section of 2.17
metres span and 0.91 metres chord.  Drag measurements were made in a wind
tunnel, and these proved invaluable in determining the optimum type of wing
covering to be used to keep the boundary layer laminar over a large part of
the section.  The structure which proved best had a combination of balsa wood
sheet and nylon fabric as a skin.  The former was used to cover the region
around the leading edge, and the remaining surface was wrapped with parachute
nylon, weighing only 25 $g/m^2$.  Owing to the effect of humidity on the tightness
of the nylon, it was found essential to apply at least three coats of cellulose
dope to ensure a consistently smooth surface.

The internal frame of the wing was made of balsa and spruce.  Two spars
were used, as this simplified the problem of jigging and incorporating the
required wing twist.  The spars were of laminated spruce.  They were joined
by cross-bracing to form a torque box, resulting in a structure which was
extremely stiff in torsion.  All ribs were of the girder type, made with
spruce and balsa.  A novel technique was used in fabricating the ailerons.
They were formed as an integral part of the main wing structure, and then the
plan-form representing the aileron was cut out and trimmed as necessary.  This
ensured excellent fit with the wing section profile.  Fin and tailplane
structure were similar to those of the wing.

Access to the control position in the box section fuselage was by means of
a removable nose section, and the main strength member supporting the pilot seat,
transmission and aircraft body was made of light alloy.  This is visible in
*Fig. 95*.  The pilot transmitted power initially to a conventional racing cycle
wheel of 0.685 metres diameter, driven by means of pedal cranks and chain.
This wheel, which provided most of the initial thrust for take-off, was in turn
connected to the propeller shaft by means of a tensioned steel belt.  Flight
testing revealed that this belt was not ideal, for although it offered strength
and weight advantages over a chain, slip occurred and occasional retensioning
of the belt was required.  It was proposed to use a wire rope system, but
although successful bench tests were completed, lack of time prevented its
incorporation in the aircraft.

The effect of including a ground wheel in the transmission to the propeller

is of interest.  Take-off distances using propeller alone, and those using a
combination of ground wheel and propeller are compared in the diagram,
*(Fig. 96).*  The advantages of a driven wheel, common to all man-powered aircraft

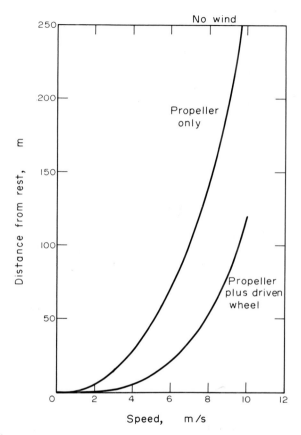

*Fig. 96* Ground runs required with and without the
assistance of a driven wheel (SUMPAC)

of this type, are most pronounced during the early stage of the acceleration
up to take-off speed, and in fact some of the aircraft would have had great
difficulty in becoming airborne under propeller thrust alone.  The indirect
drive system used by the Southampton group was found to be very sensitive to
the ground wheel to propeller gear ratio.  Too high a ratio caused excessive
power requirements for take-off, and low ratios gave the pilot uncomfortable
moments when wheel slip occurred just prior to the craft lifting off the ground.

The optimum pedalling speed was found to be of the order of 60 rev/min, and
measurements carried out at the University on the power output of a cyclist in
a reclining position agreed with the comprehensive data of Wilkie and Nonweiler.
Cruise propeller rotational speed was 240 rev/min, the aircraft having a
2.44 metre diameter propeller of conventional section with a high geometrical

pitch.

Mention should be made of the controls and instrumentation provided for
the pilot, as these generally vary widely.  The control column was mounted
directly in front of the pilot, and supported one transverse bar moved by both
hands, access being through two small holes in the front spar web.  The motions
required to operate the three control surface systems, ailerons, elevator and
rudder, are visible in the small drawing in *Fig. 92*.  No direct aids were
available to enable the pilot to judge altitude and orientation of the craft.
Small air inlet and exhaust ports were provided to maintain the free circulation
of fresh air in the cockpit.

The flight test programme lasted slightly over one year, and Derek Piggott,
the pilot on the SUMPAC first flight, was also at the controls for most of the
early test flights.  Williams, who was responsible for much of the design work
on the aircraft, also flew a few short hops.  All flights were made at the
Lasham Gliding Centre.  Very comprehensive data is available on the experience
of the pilots and the comments of the design team on the handling and design
characteristics of the aircraft.  Their opinions have proved invaluable to other
man-powered aircraft groups.

With practice it was possible to reach the take-off velocity of 9.14 m/sec
in a distance of 90 metres.  Climb rate to the cruising altitude, ideally
between 3 and 5 metres, was slow, but by holding the aircraft on the ground
until a speed in excess of the above was achieved, the climb rate could be
substantially improved.  In order to familiarise the pilot with the behaviour
of the aircraft, a number of flights were made under tow.

Piggott, whose power output would be equivalent to that of an 'average fit'
man in Wilkie's classification, achieved a maximum flight distance of 622 metres
at an altitude of 3.65 metres.  In the opinion of the pilot, he would have been
limited to flights of no greater than 1.25 km in the Southampton aircraft.  A
professional athlete should, of course, have been able to substantially increase
the range.

Turning flight was attempted on several occasions, and in some cases a
0.45 kW motor of the type used on model aircraft was used to allow increased
pilot concentration on performing the manoeuvres.  Turns of ninety degrees with
a radius of about one hundred metres were accomplished with motor assistance.

Flight testing showed that the fuselage lacked torsional stiffness, and
film of the flights shows the effect this had on the fuselage skin, severe
rippling being very noticeable.  The phugoid type of oscillation in which the
aircraft pitches about the horizontal at very low frequency, was also evident,
although no great importance has been attached to this in the write-ups of the
trials.

Drag characteristics could have been improved by reducing the wing thickness
to chord ratio, and it was also suggested that twin propellers in the wing,
similar to the layout of the Bossi-Bonomi aircraft, would have removed the
undesirable pylon interference effects.  It was believed that the acceleration
of the flow over the wings so produced would assist the stability of the
laminar boundary layer, but this is rather dubious, and the necessarily complex
transmission required for the two propellers may outweigh the small initial
drag advantage, as ground clearance and hence frontal area would have to be
increased.

An interesting feature of the structures of such light aircraft is their
capacity for water absorption, and general behaviour in humid environments.
It is quite feasible that the flight lengths of the SUMPAC could have been
substantially increased if the structure had been dry and completely free from
warping.  The humidity is thus capable of increasing both the weight and profile
drag of the aircraft.  Slackening of the nylon fabric covering also occurred as
a result of dampness, and the designers recommended a change to 'Melinex'
plastic film[1].  An estimated nine per cent saving in power would have been
possible if this re-skinning had taken place, and the aircraft had been kept
completely dry.

The undercarriage and nose section both suffered considerable damage in
crashes, and the undercarriage also collapsed during taxiing trials.  Failures
of the tyre on the drive wheel due to impact loading during touch-down were
eliminated by the adoption of a nylon-reinforced tyre.

The Southampton University aircraft, now on permanent display at Old Warden
airfield, the home of the Shuttleworth Trust, well deserves its place in the
comparatively recent record of achievements in man-powered flight in Britain,
in spite of, or perhaps due to, its unsophisticated and conventional design.
As a product of a group of young and inexperienced aeronautical engineers with
very limited resources, it was a very creditable and significant step forward.
(Fig. 97).

The Hatfield Club's machine, 'Puffin', the first flight of which took place
one week later than that of the Southampton aircraft, was the end product of a
team of professionals in many disciplines, and will be seen to be such.  The
improvement in performance was not, however, very significant.

Recommendations made by the Southampton Group for improving their aircraft
were not implemented at Southampton, but the aircraft was later moved to
Imperial College, London, where Alan Lassière, a founder member of the
Southampton Group, carried out further research work.  The abortive attempt to
rebuild the aircraft under the direction of a group at Imperial College is
detailed in Chapter 11.

* * *

---

[1] 'Melinex' is an ICI trade-name.

*Fig. 90.* An early short flight of the Southampton
University man-powered aircraft, the first
machine to take off unaided and fly successfully in the United Kingdom.

*Fig. 94.* SUMPAC being prepared for a flight attempt.
The shape of the upper wing profile is
visible.

*Fig. 95.* SUMPAC cockpit.

*Fig. 97.* SUMPAC on display following its early flights.

# 10

# 'Puffin' and her Contemporaries

## — a Professional Approach

# 10

Two quite substantial development programmes culminating in prototype man-powered aircraft were commenced in Britain before the Southampton University aircraft was conceived.  These machines were the Perkins inflatable wing aircraft, and the Hartman ornithopter.

To describe the activities surrounding the design, construction and trials of these aircraft, it is necessary to temporarily divert from the linked activities described in the previous two chapters, as these two machines were not conceived with the Kremer Competition in mind, and several of the features incorporated in the construction remain unique as far as man-powered aircraft are concerned.

Although neither of these prototypes was as successful as the Southampton aircraft, if indeed longevity of flight was the sole aim or even primary aim of their designers, both made a contribution to the science of man-powered flight.  It is likely that the concept of Perkins' aircraft may have wider applications outside the field of immediate interest.

## DANIEL PERKINS' 'PNEUMATIC' AIRCRAFT

A full report on the work by the late Daniel Perkins, who was a Senior Experimental Officer with the Ministry of Defence at the Research and Development Establishment, Cardington, was published in 1963.  The report describes in detail the construction and trials of two prototypes of the 'pneumatic' man-powered aircraft.  The results will be dealt with in some detail here because, as the author states in his summary:  "... over the speed range covered, the efficiency of a propeller is very adversely affected when the machine is in motion, by the presence of the ground".  Use of the inflatable wing enabled exceptional lift coefficients to be obtained, only bettered by rotating cylinders, (see Chapter 15).

The decision to adopt inflated wings was made following a simple study of the requirement of a man-powered aircraft.  The designers argued that with a continuous input of 0.38 kW, the aircraft must be large enough to fly at 38 km/h with an all-up weight of 91 kg.  Assuming a propeller efficiency of 75 per cent, some 285 J[1] of energy would be available from the propeller per second.  Limiting the total aircraft drag to 3.18 kg, this meant that the machine would need an overall lift to drag ratio of at least 28.6.  Assuming that the pilot weighed 68 kg, the machine would have to weigh no more than 23 kg.  The design team considered it unlikely that these requirements could be satisfied, using conventional materials and fabrication techniques.

---

[1] *A Joule (J) is the work done when the point of application of a force equal to to one Newton is displaced through a distance of one metre in the direction of the force (1 Joule = 0.738 ft.lb)*

Previous work at Cardington on powered aircraft with inflated wings had
shown that light-weight robust aerofoils could be made to operate at high lift
to drag ratios. (Spans and aspect ratios had, until the man-powered project,
been very low, however). Because of this lack of information of high aspect
ratio inflated wings, the first task was to test the design principle on units
with aspect ratios of fifteen and thickness to chord ratios of the order of
twenty-two per cent. A wing with these characteristics, and an area of 9.29 $m^2$
was successfully tested with single wire bracing and found to be sufficiently
strong to provide a factor of safety of 1.5. The total wing weight was only
10 kg. Using this as a basis for the weight estimate of a complete man-powered
aircraft, Perkins and Lock produced a first estimate of the empty weight as
being 28.1 kg. Taking published data on power outputs, (their potential pilot
was thought capable of equalling the figures published), an aircraft with an
all-up weight of 88.5 kg (pilot weight 60.4 kg) could be produced. This was
within the first estimates assumed for the minimum weight.

Before commencing the construction of a prototype, a three-wheeled test
vehicle was made in order to explore the potential efficiency of the propeller
and associated transmission gear. The most satisfactory drive mechanism was
found to be a rubber proofed rope belt running over Vee pulleys. Primary
drive was through pedals, the pilot in a semi-reclining position.

The 2.9 metre propeller was designed to produce maximum efficiency at
32 km/h. Blades were made up using 1 mm thick birch ply around a strengthening
spar of dural tube. Automatic pitch correction, a novelty on man-powered aircraft, was implemented using the centrifugal force set up by the blade as it
rotated to proportionally extend a spring controlling the blade angle. The
static rotational speed was 240 rev/min, producing approximately 135 kg thrust.

During vehicle trials on the runway, maximum speeds of 22.5 km/h were
reached, with the trolley covering the 27.5 metre test run from rest in eight
seconds. The results were considered satisfactory because no streamlining had
been carried out, the pilot sitting in the open air; also the vehicle
weighed 109 kg more than the proposed aircraft, and had been made out of scrap
metal. Encouragingly, the propeller weight, including the variable pitch
control mechanism, was 3.63 kg less than the design estimate. In total, the
above evidence suggested that the construction of the prototype should still
proceed.

Construction of what was to be the first prototype commenced in February,
1959. The wing area was maintained around 9.3 $m^2$, and the wing was of
symmetrical section with an aspect ratio of 15, a thickness to chord ratio of
20 per cent and a 3:1 taper ratio. The wing was mounted on struts above the
pod fuselage, sitting on a tricycle wheel arrangement. It was controlled by
varying the wing incidence, in addition to a conventional rudder. Final AUW
was 100 kg, ten per cent above the earlier quoted maximum value.

It was during acceleration runs in a hangar at Cardington that the first
hints, at this stage unsuspected, of the propeller performance being adversely
affected by the ground proximity appeared. The acceleration runs and speed
trials revealed performances no better than that of the trolley used to test
the propeller! This was despite the fact that the former was lighter (the
wings had been removed, and the fuselage unit weighed 88.5 kg including the
pilot), and was also highly streamlined.

A complex series of trials were started in an attempt to find the explanation for these results.  During this period other seemingly inconsistent results were obtained:

1.  The selection of gear ratio and pitch setting had no effect on the top speed.

2.  Increasing the number of propeller blades in steps up to five, and allowing the pilot to manually vary pitch, generally resulted in a decrease in performance.

3.  Towing trials on the runway indicated that only about 0.38 kW was needed to achieve take-off.  The pilot pedalling rate during free trials suggested that 0.60 kW was being used with no take-off occurring.

4.  Measurements on the power output of the pilot showed that 0.60 kW was being generated for the initial 30 seconds, and these measurements were remarkably consistent to within 5 per cent.

5.  Drive system efficiency was found to be 91.3 per cent - this was quite acceptable.

6.  Wheel friction was normal during the ground run.

7.  Fuselage drag was low.  No part of the machine was swept by the propeller slipstream, therefore no turbulent drag would be encountered.

The general concensus of technical opinion of those expert in propeller performance was that there should have been no decrease in efficiency due to the presence of the ground.  It seemed possible to the designers that the fuselage drag could have been increased when the propeller was being used, due to evidence of point 4, above, and it was decided to reduce fuselage drag, although this involved a complete aircraft redesign and rebuild.

## TESTS ON THE REDESIGNED AIRCRAFT

The first prototype of the Perkins machine had a frontal area of 1.39 $m^2$ excluding the wing.  The re-designed model, a conventional low wing tractor monoplane had a corresponding area of only 0.465 $m^2$, and a reduced empty weight of 26.8 kg.  This very low frontal area was achieved by reclining the pilot and arranging his line of sight such that a canopy could be made redundant, instead providing for vision through a transparent propeller spinner.  (This of course, meant that the complete fuselage would be in the turbulent wake of the propeller; however, the designers felt that the reduction in fuselage cross-section would far outweigh this disadvantage).  The original wing and the best gearing and propeller found during tests on the first prototype were retained, *(Fig. 98)*.

Prior to test runs on the ground, an air speed indicator (ASI) was attached to the wing tip, well out of the way of the propeller slipstream.  Ground runs under pilot-power alone showed a top speed of only 21.5 km/h, similar to the top speed of the first prototype.  The designers were faced with the baffling fact that three vehicles, all with widely differing geometrical features, all had a maximum speed of approximately 22 km/h.

The next scheme adopted in an attempt to improve the ground speed was the use of a flapping propeller. (Apparently increased efficiency is experienced in helicopter rotor blades in translational flight, as a result of undisturbed air being passed through the blade 'disc'. This compares with the hovering and vertical flight modes, where a pre-slipstream is formed. A flapping propeller is able to take advantage of a simulated translational mode.) This was tested on another vehicle in the hangar, and a ground speed of 22.5 km/h was achieved in a distance of 45 metres. Further experiments with automatic pitch adjustment yielded a top speed of 28 km/h.

On fitting the unit to the second prototype aircraft and carrying out a man-powered trial, a maximum speed of only 19.3 km/h was reached. A complete review of the project was then undertaken. While the first prototype had not fully met design specifications, the second machine had been lighter than originally estimated, high lift coefficients had been obtained, and the power output of the pilot was up to scratch. There appeared to be only one absolutely common factor to the four vehicles constructed - the proximity of the propeller to the ground.

The first check on this theory was to show that almost any body generated lift near the ground, regardless of its lift-producing capabilities at altitude. Tests were made using a 1.83 metre long cylinder, weighing 0.227 kg, winched at speeds of up to 32 km/h. When these experiments were carried out close to the ground, the cylinder 'flew' every time at heights of up to 1.8 metres[1].

Perkins then performed a clever little experiment which was most revealing. Two card discs were mounted, one at each end, on a 3.6 metre long pole suspended at the centre by a thin thread. After statically and dynamically balancing the whole unit, the pole was suspended horizontally, the discs facing the direction of motion. The pole was then moved forward. The pole did not rotate and it was concluded that the drag of each disc was equal.

The pole was then mounted vertically, with one disc as close as possible to the ground. On moving the unit forward, the lower end moved back "... with sufficient velocity to make the pole complete the circle".

Checks were made to ensure that this was not an isolated reaction, and it was concluded that adverse effects on propellers close to the ground could readily occur. The pattern of flow around the propeller was thought to be restricted to the top and sides by the ground, this disturbance affecting the thrust characteristics.

Experiments to ascertain to what extent the dynamic efficiency of the propeller could be affected by the closeness of the ground suggested losses in performance of up to thirty-three per cent.

In concluding their report on the trials, the authors did feel that the pneumatic aerofoil, and its encouraging performance near the ground "... appeared to offer an excellent basis on which to design a small, cheap and relatively low-powered ground effect machine".

They were not unaware of the fact that the ground effect on propellers

---

[1] *This phenomenon is not to be confused with the Magnus effect, which may only be demonstrated with a rotating cylinder (or similar body). Reference to this is made in Chapter 15, and it is defined in Appendix VI.*

could occur on large powered aircraft.  Perkins stated:"

> "... most aircraft have aerofoils not sufficiently asymmetric to preclude
> the possibility of generating some pressure lift, and bearing in mind the
> inaccuracy of ASI pitot heads at their wrong altitude, there seems quite
> a possibility that conventional aircraft achieve take-off at speeds lower
> than calculated.  This would tend to hide any loss in the efficiency of
> the propeller.
>
> Equally, if an aircraft is subjected to overloading, or adverse runway
> conditions such as deep slush, it may also develop the same symptoms
> as the man-powered machine."

Almost four years later, in July 1966, David Rendel visited Cardington to
view the third man-powered aircraft constructed by Perkins, *(see Fig. 99)*.
This machine resembled a flying wing, the platform being triangular with a
central cut-out for the pilot.  Fixed stabilising fins were located on each
wing tip, and a single all moving fin was sited behind the pilot.  The propeller
(still close to the ground!) was mounted on this fin in such a way that it also
moved as rudder control was effected.  The wing had a very low aspect ratio, 4,
but the area of 23.4 m² was large compared with earlier prototypes.  With a
span of 8.23 metres, the wing weighed only 5.9 kg (total empty weight of the
machine being 17.3 kg).

Structurally simple, the main feature apart from the inflated wing was a
tapered tube running fore and aft at the wing centre section.  This tube
carried a steerable nosewheel, the pilot seat and transmission mechanism,
controls and fin support pillar.  The pilot was reclined such that his pedals
were at approximately the same height as the saddle.  A horizontal bar running
beneath his knees was twisted for pitch control, steered as handlebars for yaw,
and rocked for bank.  Control surfaces were built into the wing.

Propeller diameter was 2.58 metres, and a continuous belt drive was used
to rotate it at 240 rev/min.  As on Perkins' earlier machines, the main ground
wheel was not driven.

Claimed take-off speed was 24 km/h, more in line with the maximum ground
speeds attained in the previous experiments, with a cruising speed of 29 km/h.
Flights of up to 128 metres were made in the large Cardington hangar, and under
man-power alone, an independent witness saw the machine climb to an altitude
of about 0.6 metres.  A large number of towed take-offs were also made.

Daniel Perkins relied solely on his own finance for the construction of his
aircraft, which were made in his spare time.  The delta wing machine, for
example, took 800 hours of labour to manufacture, and cost £200.  For a
designer whose background was textile engineering, Perkins introduced many
novel and sound ideas associated with inflatable aircraft.  One of these was a
pneumatically operated aileron, the 'spring-loaded' effect of the trailing edge
being used to return elevators to their equilibrium position.

Although Perkins did not propose manufacturing further prototypes, he
suggested that a delta wing machine with a 12.2 metre wing span would be a
suitable size for longer flights, and with such a machine Perkins was confident
that the Kremer course could be conquered.

## THE EMIEL HARTMAN ORNITHOPTER

The Hartman man-powered ornithopter was of totally different concept to the Perkins machine.  Emiel Hartman, a London sculptor, began the design of his aircraft in February, 1958, seemingly against the advice of the majority of members of MAPAC, who were doubtful of the feasibility of flapping flight.

Hartman's machine[1] was of 10.96 metres span and was built using the then conventional sailplane techniques and materials, these being wood and fabric, (see Fig. 100).  Normal tail surfaces were incorporated, and a tricycle undercarriage with steerable nosewheel was used.  The wing was hinged at the root and also on the single spar.  Pinions on the trailing edge and tips of the wings enabled the wing incidence in effect to be varied, it being necessary to increase the incidence on the upstroke, and reduce it during the downstroke. (A similar device was used by Filter in Germany in 1958.  His machine was shown at the Hanover exhibition of that year).

Restraints were imposed to limit the upward and downward movement of the wings, and rubber strands were so arranged as to keep the wings in a near-horizontal position when unloaded.  The pilot used his arms and legs to provide the power input, most of the effort being put into the leg movements. Forward thrust of the feet, either together or independently, pulled cables which passed outside the fuselage to fixtures on the wing.  This resulted in a downward motion of the wing.  In conjunction with the thrust forward of the feet, the pilot pushed the diamond-shaped structure (visible on the side of the fuselage) forwards using his hands.  This aided the downstroke of the wings. On easing back his feet, the pilot allowed the tension in the rubber cables to reduce, lifting the wings to their highest position.  If the handgrips were simultaneously pulled back, the process was accelerated.  It was Hartman's intention to operate the wings at a flapping rate corresponding to its natural frequency, thus considerably easing the effort involved in moving them.

Rendel, then working at the Royal Aircraft Establishment, Farnborough, commenting on Hartman's machine in the magazine *Flight*, in October 1959, lists several advantages of ornithopters, as follows:

1.  Propulsive efficiencies of eighty to ninety per cent can be achieved
    if forward speeds are kept low and the wings correctly twisted.

2.  Oscillation of the wings can delay the build-up of induced drag.

3.  Boundary layer control possible by birds because of permeable wings
    can greatly increase the lifting power.

4.  An elastic wing structure could adapt to changing conditions of flight.

He continues:

"Hartman's aircraft appears to have non-permeable in-elastic wings, and so it will not be able to benefit from many of the advantages described above. Furthermore, it has apparently no built-in twisting mechanism, and therefore may not achieve the high values of propulsive efficiency which are

[1] *Some of the features of the Hartman aircraft are the subject of British Patent No. 851,352.*

theoretically possible."

Rendel does have some praise, however, for the flapping system on this aircraft:

"One of the features of the aircraft ... is the incorporation of an
elastic suspension system, whereby the wing oscillates in resonance with
the natural frequency of stretched bungee cords adjusted to conform to the
natural frequency of the human body when in the rowing attitude".

He later states:

"For man the aerodynamically optimum flapping speed comes out at between
thirty and forty degrees per second, or roughly one half flap per second
when in level flight. This conforms to the rate of natural movement when
rowing ... and also to the elastic support system built into Hartman's
aircraft. Thus the man, the machine and the air are, as you might say,
all in resonance with each other – and this should lead to considerably
more efficient power utilisation. It may also be found that better power
production is possible in this way."

Towing trials were carried out at the College of Aeronautics, Cranfield,
the aircraft being pulled behind a car at speeds of up to 64 km/h during which
time short hops were made. No free flights were made.

At an R.Ae.S. Man-Powered Aircraft Group Committee meeting on 6th January,
1961, members discussed a letter from Emiel Hartman requesting a grant to
enable him to carry on work on the machine. Several modifications were
required, but no money was forthcoming.

(The same meeting reaffirmed their interest in the work of Perkins at
Cardington, and requested progress reports. It also considered making Group
funds available to finance the construction of a new propeller for the machine.)

## THE 'PUFFIN' AIRCRAFT

Considerable credit has been given in Chapter 9 to the team which designed
and constructed the Southampton University man-powered aircraft. Similar credit
should be attributed to the group at Hatfield who constructed 'Puffin', and
later 'Puffin II'. This team was formed of professional aeronautical engineers
with much experience in the design of aircraft such as the Trident – a far cry
from the concept of man-powered craft.

However, the Hatfield group were offered facilities unavailable to the
students at Southampton. Also the Southampton University team were largely
influenced by the term of their undergraduate and postgraduate research
programmes. These normally do not exceed three years.

The first flight of the 'Puffin' man-powered aircraft occurred one week
later than that of its main competitor at Southampton.

The 'Puffin', a most apt name for the aircraft, was, like the Southampton
machine, the product of a sizeable group of enthusiasts, the majority of whom
were working at what was then the de Havilland Aircraft Company, Hatfield.
John Wimpenny (see Fig. 101), the Chairman of the Hatfield Man-Powered Aircraft
Club, and an aerodynamicist by profession, had been interested in the prospects

for successful man-powered flight since the mid 1940s, having constructed a model to demonstrate the essential features of this type of aircraft in 1946. Wimpenny's work formed the basis of the HMPAC machine design, and a photograph of the prototype is shown in *Fig. 102*. (A second aircraft, 'Puffin II', was built by the Hatfield group, following an accident to the first prototype in 1963).

Main features of the aircraft are the large wing span, tail mounted pusher propeller, and the pilot position. The design philosophy was by necessity similar to that followed by the Southampton team, being influenced by the Kremer Competition. Also, the conventional layout of the flying surfaces, utilising a single wing of high aspect ratio, was considered the optimum form on which to base an aircraft which could be flown by an average fit man.

By locating the propeller on the rudder, the additional drag of a pylon was removed, but the transmission of power to the propeller had to be by means of a comparatively unorthodox process. Features of the transmission system are shown in the *Flight* drawing *(Fig. 103)*.

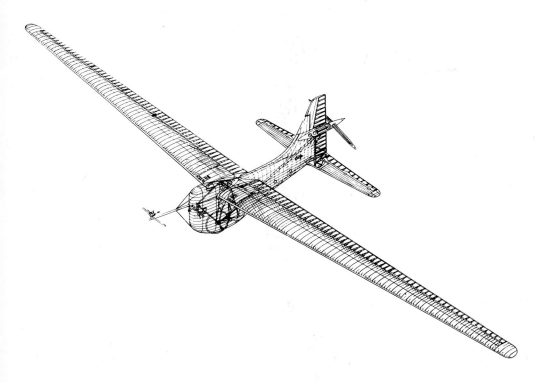

*Fig. 103*  The HMPAC 'Puffin', first of two aircraft constructed at Hatfield. The torque tube for transmission of power is visible.

Conventional cycle pedals drove the wheel and propeller simultaneously.
A torque-shaft bevel gear transmission to the propeller was used, and a
coaxial gear drive between the pedal cranks and the ground wheel provided take-
off acceleration.  The drive shaft to the propeller was initially constructed
using balsa wood;  however, on one test flight the shaft broke during the
initial climb manoeuvre, and was subsequently replaced by a metal tube.
Transmission gearing was of steel, but magnesium tube was used for the support
frames.  Unlike the Southampton aircraft, no intentional suspension gear was
incorporated for cushioning landing impacts, the only shock absorption being
the tyres on the drive wheel.  Much of the credit for the transmission system
and drive must be given to the Dunlop Company (Aviation Division) who undertook
all detail development and manufacturing work.

At its peak, the Hatfield Man-Powered Aircraft Club consisted of approximately forty full members, a team of apprentices from the de Havilland
Aeronautical Technical School and a number of honorary members.  Most of the
team were employed at Hatfield, and most of the disciplines found within the
structure of a major aircraft company were represented.

The impetus leading to the formation of this team was given by the
announcement of the Kremer Competition, and the HMPAC was officially formed on
12th August, 1960, with the blessing of the de Havilland directors.  Initial
financing of the project was provided by the founder-members, including
J.C. Wimpenny, E.C. Clear-Hill, S.C. Caliendi, C.J. Goodwin, J.W. Haggas,
D.R. Newman and F.W. Vann.

Having completed a programme of structural tests and investigations into
the optimum pilot position, a submission was made to the Royal Aeronautical
Society, in the hope of obtaining a grant similar to that received by the
Southampton Group.  In spite of many offers of support, mainly in the form of
material, by firms donating items ranging from Sellotape to the total amount
of balsa wood required for construction, the club members needed a further
£1500 to enable the project to proceed to completion;  this sum was given out
of the fund set up by the Man-Powered Aircraft Group of the Royal Aeronautical
Society.  Notification of the award was made on 27th February, 1961.

It must be emphasised here that awards were not given without close
scrutiny of the technical and economic merits of the proposals.  Rendel, in
comparing the Hodgess-Roper design (Chapter 11) with 'Puffin', favoured the
former because of its shorter wing span, making for easier handling, and his
seemingly 'more realistic' weight estimate, (an AUW of 136.3 kg compared with
'Puffin's' predicted 97.6 kg, in fact the latter's typical AUW was of the order
of 120.5 kg).  There was also some criticism that the Hatfield project was
being costed on the basis of aircraft industry practice!

Lilley, also a member of the Man-Powered Aircraft Group Committee, had
misgivings regarding the 'Puffin' wing weight.  If the wing was light, it would
probably suffer from inadequate torsional stiffness, affecting boundary layer
transition and hence drag.  He also felt that the propeller was situated too
close to the ground, no allowance being made for the loss of thrust during
take-off.  (This may have been prompted by the experiences of Perkins).

A full Committee Meeting in January, 1961, asked HMPAC to give consideration
to the points raised by Lilley, and following discussions the grant was approved

Most of the construction work was carried out by the apprentices of the

D.H. Technical School, assisted by three retired wood-work craftsmen;  a
model aircraft club also assisted.  The mating of the various components took
place in the de Havilland Engine Company hangar in August, 1961.

In parallel with the construction of the airframe, a large amount of
development was being put into the design and testing of the propulsion system.
A very versatile rig was used to ascertain the optimum pilot position, from
the point of view of both maximum power output and comfortable posture.  A
bicycle frame formed the basis of the rig, on which was also mounted the
propeller.  The pilot operating the rig was able to read the thrust power
directly using a dial situated in front of him, and artificial 'feel' was
introduced at a later stage to enable familiarity with control forces and
amplitudes to be obtained.

A flight simulator was used for practicing roll control, and the Hatfield
pilots stressed that the control of a man-powered aircraft in the roll mode was
unlike that of a glider or powered light aircraft.

It was found that the initiation of roll was relatively easy, but problems
were met once an attempt was made to stop the roll.  Very large ailerons, having
large deflection angles, were needed to counter the possibility of turning into
the ground.  Aileron deflections of seventy degrees were used, this increasing
the wing drag locally by a considerable amount.

The first wheel-out of 'Puffin' occurred at 10.15 p.m. on 15th November,
1961, and was the occasion of a small naming ceremony.  Taxiing trials were
commenced without delay, John Wimpenny carrying out the first of these.
Jimmy Phillips piloted the first flight, which took place on 16th November,
using the main Hatfield runway.  The aircraft was pointed into the 5 - 6 km/h
wind and the static aileron response was checked.  A slow taxi run was made to
verify the rudder and aileron feel at higher speeds, and the speed was
gradually increased up to take-off speed.  Within two days, flights of up to
one minute in duration were being made, *(see Fig. 104)*.

Early problems to be overcome were discovered during taxiing up to take-off
speed.  The original tail skid was replaced by a small wheel because the former
was found to have a significant drag penalty.  It was noticed that the elevator
motion was insufficient to raise the tail to the correct altitude during the
acceleration up to take-off speed.  Also the rate of pedalling increased as the
aircraft left the ground, and in order to ensure a relatively smooth transition
between take-off and flight, the propeller pitch was adjusted.

Other minor problems affecting pilot performance during the early tests were
oxygen starvation due to the very small air inlet duct in the cockpit, and lack
of instrumentation, an air speed indicator being the sole flying aid.  The
flying characteristics were generally satisfactory but the elevator was over-
geared, tending to result in over-corrected pitch control.  Although the slow
response of the aircraft to changes in roll and yaw control was a hazard, the
controls appeared effective if deflections were held for some time.  One
interesting result of the pilot position adopted,  in which the pilot's head
was 0.76 metres from the ground, was the gross over-estimation of altitude
resulting from the fact that any increase above 0.76 metres seemed large in
proportion.

As the pilots gained confidence in controlling the aircraft, gentle turns
and weaving flights were attempted.  The most important factor delaying the

progress in making the large turns necessary in the Kremer figure-of-eight course was the short flight duration. Attempts to increase flight time were made by towing the aircraft behind a car. However, this was soon aborted as instabilities occurred.

Following a series of successful flights over long distances, the maximum being 911 metres, (variously reported as 908 metres)[1] a 0.19 kW model aircraft engine was attached to the side of the fuselage, in a final attempt to obtain sufficient power to remain airborne and complete a series of prolonged turns. At the second A.G.M. of the Man-Powered Aircraft Group, on 11th May, 1962, the Chairman was able to report that upwards of fifty flights had been made. A 4.5 metre altitude had been reached, and the machine handled well, being very stable. With a pilot weight of 66.8 kg the average flying weight was 116.6 kg. Prior to use of the engine, turns of 83 degrees had been accomplished. This proved a useful aid, and the aircraft was able to take-off, complete a 270 degre turn to starboard, and a 90 degree turn to port, landing in the opposite direction to take-off. John Wimpenny, who carried out most of the longer flights, commented that control was not a great drawback in manoeuvring the aircraft, provided that the pilot was able to conserve sufficient reserves of power.

One of the most valuable pieces of data accruing to the flying experience of pilots in 'Puffin' was the need for a visible horizon. On one day when conditions were misty, the aircraft crashed and six metres of one wing were destroyed. The reason for the crash was attributed to the inability of the pilot to judge his altitude with regard to a fixed horizon.

Several flights were made in 'Puffin' by Chris Church, a racing cyclist. He was at the controls when, early in April, 1963, 'Puffin' was seriously damaged in a flying accident. The aircraft was flying down the Hatfield runway when a side gust blew it off course. Before corrective measures could be taken the aircraft touched down in mud off the side of the runway. This brought 'Puffin' to an abrupt halt, and the wings folded forward. The official assessment of the damage revealed that the fuselage would require completely rebuilding ahead of the tail portion. Also, the wing, it was thought, could be repaired, but the effort involved was likely to be as great as starting from scratch and constructing a new wing. The empennage, propeller, gearbox and miscellaneous small fittings could be used again.

This demise did not mean the end of the Hatfield Group. On the contrary, the accident was used to its full advantage in that a major re-design of 'Puffin' was made, largely based on flying experience in the first prototype. The 'Puffin' had also been insured for £4500!

The result of the re-design was 'Puffin II', (see Fig. 105), an aircraft with a wing span of 28.3 metres, 2.74 metres greater than that of the original 'Puffin'. The corresponding flying weight of 116 kg was almost identical to that of its predecessor, and the wing loading was thus considerably reduced. A new wing section was used and instrumentation was improved, there now being incorporated a pitch and bank indicator, the latter showing the maximum

---

[1] *Wimpenny was awarded a prize of £50 for this flight, the first of over one half mile (0.805 km) solely under man-power. An anonymous donor was responsib for the gift. This flight was witnessed by B.S. Shenstone and also by representatives from thirteen countries attending an IATA technical meeting in Londo*

permissible bank angle, as well as the ASI. One significant outcome of the new
wing design was that the power required to fly in ground effect was reduced
from 0.30 kW to 0.225 kW. Aspect ratio was marginally increased from 21.4 to
22.

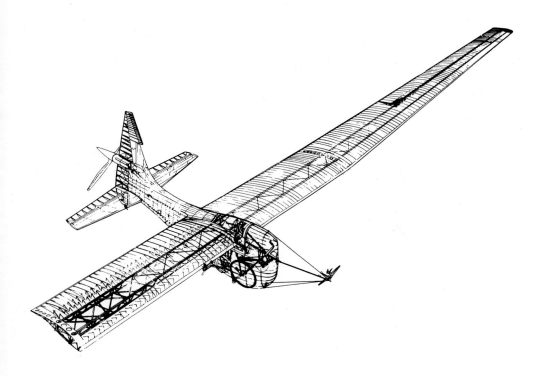

*Fig. 105* Puffin II. Note the different wing plan,
compared with the constant taper of the first prototype.

'Puffin II' completed its first flight on 27th August 1965 at 6.30 p.m.
with Phillips as pilot, *(see Fig. 106)*. Control of yaw proved a problem, and
this was not improved even when a larger rudder and tail fin were fitted, *(see
Fig. 107)*. Tests were also carried out on 'Puffin II' using an auxiliary
engine. Flying experience led to the conclusion that ailerons were not
required for control when only small turns or corrections to a straight course
were needed, and the rudder was effective in contributing to these manoeuvres,
*(see Figs. 108 and 109)*.

A number of flights in excess of half a mile (0.8 km) were carried out by
'Puffin II' in the two years up to mid-1967, but as reported at the A.G.M.s of
the Royal Aeronautical Society Man-Powered Aircraft Group for the years 1966

and 1967, HMPAC, in common with other groups, had been experiencing delays in
flight programmes caused by unfavourable weather, and also by lack of pilots.

However, in August, 1967, 'Puffin II', having achieved a 5.2 metre altitude
and flown 0.8 km, clearing a 3.05 metre high obstacle at the end of the flight,
made preliminary attempts on a 'slalom' course, *(see Figs. 110 and 111)*.

## A NEW KREMER COMPETITION — THE 'SLALOM' COURSE

This 'slalom' course was an additional incentive, announced on 1st March,
1967 by the Royal Aeronautical Society.  Performance of man-powered aircraft in
the previous six years had clearly been disappointing, not least to Henry Kremer
who had given the projects such magnanimous support in the early 1960s.

It was evident from the tones of the discussion at each successive A.G.M.
of the Man-Powered Aircraft Group that a continuing optimism had prevailed.
In May 1964, for example, at the fourth Annual General Meeting of the Group,
it was stated that four aircraft, those at Southend, London University (ex-
SUMPAC), Woodford and Hatfield ('Puffin II') were all likely to fly later in
1964, and it was agreed at this meeting that the Group should continue
functioning to further the interest in man-powered flight after the Kremer
Competition had been won.  In fact, none of the above aircraft flew in 1964,
and only one, 'Puffin II' flew in 1965, as described earlier in this chapter.

By now, of course, the original limit set as the final date for winning
the Kremer Competition, February 1962, had long been passed, and little
progress had been made towards the completion of the Kremer course since the
early flights of 'Puffin' and the Southampton University man-powered aircraft.

This led to the introduction of a further contest which was known as the
£5000 Kremer Competition, together with the increase in prize money for
completion of the original figure-of-eight course, this contest now being
called 'The £10 000 Kremer Competition'.  A further concession was that the
£10 000 Kremer Competition would, from 1st March 1967, be open to entrants from
any country in the world, although the £5000 competition was initially
restricted to design teams from member countries of the British Commonwealth.

The additional prize money was again the generous gift of Henry Kremer,
and until the competitions are won, the interest from the prize money invested
provides considerable amounts for grants which are given to the more promising
potential competitors.

In order to ensure uniformity of regulations in the world-wide Kremer
Competition, The Royal Aero Club competition rules were supplemented by the
Sporting Code of the Federation Aeronautique Internationale, (FAI) and it was
suggested that actual attempts on the Kremer course should preferably be made
in countries affiliated to the FAI.  The regulations remain unchanged, but the
Royal Aeronautical Society reserve the right to review and amend these
regulations should the prize still be unclaimed.

The £5000 Kremer Competition, identified by the 'slalom' course, was
introduced as a way of minimising the amount of turning required by the aircraft,
while maintaining a sufficiently difficult course to ensure that both pilot skill
and aircraft design were of a necessarily high calibre.  The prize money was to

be given to the first three successful competitors, these receiving awards of
£2500, £1500 and £1000 respectively.

Conditions of entry were similar to those for the main Kremer Competition.
The requirements appertaining to the aircraft were identical to those in the
original rules, but a rider was added to Section 4.2.3. governing crew
conditions, allowing the aircraft to be manually turned on the ground by
ground crew before the return flight (see course details) was attempted.  The
plan of a typical course is shown in *Fig. 112*.

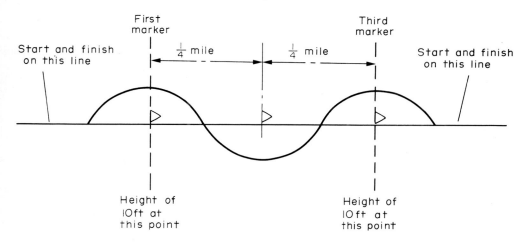

*Fig. 112*  The Kremer Competition 'slalom' course

The official rules for the competition (see Appendix V) described the
course as follows:

"The course shall consist of two flights in opposite directions, each
including three turns made around three markers spaced at intervals of a
quarter of a mile in a straight line.  The starting point shall be on the
extension of the line of the markers and may be as near to the first marker
as the competitor wishes, provided that the machine is at a height of ten
feet (3.05 metres) when it passes the first and third markers.

The course must be completed in both directions within an elapsed period
of one hour from the start.  The finishing line in each case shall be an
extension of the line of the markers.

The machine shall be flown clear of and outside each marker, with the turn
round the central marker to port in one direction and to starboard in the
other.

The height of the machine above the ground shall be not less than ten feet
(3.05 metres) when passing the first and third markers;  otherwise there
shall be no restriction as to height, but after each take-off the machine
must remain in continuous flight until the finishing line at the opposite
end of the course is crossed."

A point not made clear in the regulations was that the same crew was required to make the two flights. It was also permitted to make more than one attempt on the Kremer course on any one day.

'Puffin II' was the first aircraft to attempt a part of this course, with John Wimpenny at the controls. The attempt was abortive - difficulty was met in gaining sufficient altitude, and once the aircraft was about 3.3 metres from the ground, the pilot encountered what was thought to be a crosswind, and the aircraft then landed on the grass at the edge of the runway. It was found that the wing had contacted one of the poles marking the course, this causing the change in direction.

No further flights of significance were made in 1967, but a new pilot was recruited, joining the Hatfield team in January 1968 - John Wimwell had gliding experience and was also an amateur cyclist. The first three months were spent training on the rig; Wimwell managed to maintain a power of 0.53 kW for ninetee seconds, and several times figures of the order of 0.4 kW were achieved, althoug this was difficult to reach and, of course, even more difficult to maintain for a useful period.

One very interesting result of Wimwell's work was the conclusion that gliding experience was not necessarily an advantage when flying a man-powered aircraft. It was also felt that the 'Puffin' controls should have had the same sense as those familiar to a cyclist, especially if a racing cyclist was to be the pilot.

Flight tests commenced in July of that year and on 21st August flights of 457 and 640 metres were made at 1 - 2 metres altitude. Again the impression given to the pilot of a greater altitude was a drawback.

Early in October the aircraft was thoroughly cleaned, all dust being washed off the Melinex skin, *(see Fig. 113).* This seemed to improve the performance - it should have marginally reduced the drag - and a flight of about 800 metres was promptly made. On the 7th October, a flight was attempted to achieve a maximum distance, but 'Puffin II' only achieved 540 metres on this occasion, the average altitude being 1.83 metres. A final attempt made on 9th October was a flight of 82 metres at an altitude of 2.74 metres *(see Fig. 114).* A 4.5 km/h cross wind blew up during this flight, and its detrimental effect on performance and control was very noticeable.

Some useful lessons were the result of the series of flights carried out towards the end of 1968. Pilots found that considerable nervous energy was expended in changing focus to view alternately the horizon bar and air speed indicator. During turns it was often convenient to use the secondary effects of ailerons instead of putting full effort into use of the rudder.

Several attempts were made at taking off down-wind, but this form of procedure was not recommended to future pilots as the increased speed seemed particularly alarming! If turns were attempted after prolonged straight and level flights, they were generally associated with increases in drag sufficient to cause a loss of altitude.

The final flight of 'Puffin II' at Hatfield was made on 6th April, 1969. The aircraft completed a straight run down the perimeter track and was wheeled on to the runway and lined up for take-off. There was no sign of wind, and take-off proceeded uneventfully. Suddenly a gust of 6 - 7 km/h blew normal to

the flight path of the aircraft, and 'Puffin II' drifted across the runway.  A
touch down was made on the runway after the aircraft had yawed to the left, but
it immediately left the runway and started rolling across the grass.  The pilot
did not have sufficient speed for effective control of direction with rudder or
ailerons, and application of the hand brake was unsuccessful in slowing the
aircraft.  After rolling a short distance 'Puffin II' collided with a concrete
plinth carrying one of the runway lights, and the complete airframe collapsed.

This unfortunate end to the Hatfield Man-Powered Aircraft Group's activities
was not the last we will see of 'Puffin II', however, and its subsequent rebirth
as an entirely different 'bird' at Liverpool University is described in
Chapter 16.

* * *

*Fig. 98.* Perkins' early inflatable wing aircraft at
Cardington.

*Fig. 99.* Daniel Perkins' final man-powered aircraft
with an inflatable wing - 'Reluctant Phoenix'.

*Fig. 100.* The Hartman ornithopter.

*Fig. 101.* John Wimpenny in Puffin.

*Fig. 102.* Puffin during weighing before the first
flight.

*Fig. 104.* Puffin in flight, 1962.

*Fig. 106.* Jimmy Phillips at the controls of Puffin II.

*Fig. 107.* Puffin II, showing the fin extension. (The dihedral was later increased).

*Fig. 108.* Puffin II photographed during 1965.

*Fig. 109.* Puffin II during the first test flight,
27 August 1965.

*Fig. 110.* The centre wing structure of Puffin II.

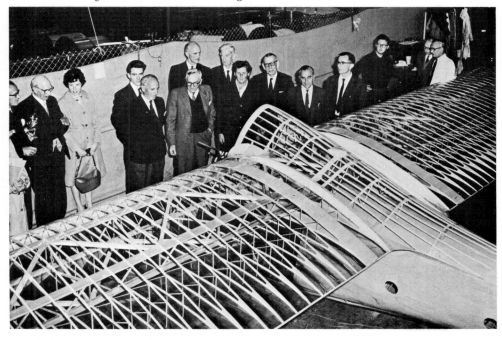

*Fig. 111.*    The Royal Aeronautical Society Man-Powered Aircraft Group inspecting progress on Puffin II during 1965. J.C. Wimpenny is 4th from right, next to Henry Kremer. Miss E.C. Pike, Secretary of the Man-Powered Aircraft Group, is in the centre of the picture.

*Fig. 113.* This photograph of Puffin II shows the high
quality of finish achieved by the Hatfield
group.

*Fig. 114.* Puffin II in flight early in 1969.   J.C.
Wimpenny was pilot, and the altitude is
approximately 4 metres.

# A United Kingdom 'Miscellany' of the 1960s

# 11

   The Southampton University aircraft and the two machines constructed at
Hatfield were remarkable because of their success.  However, it is desirable,
if not necessary in a work such as this to deal with less successful contemporary
machines, or those which only reached the design study stage.  These may be
justifiably divided into two categories - those designed by professional
engineers with sufficient knowledge and experience to contribute to the science
of man-powered flight, and the product of the amateur enthusiast whose knowledge
of aeronautical engineering and the 'science of flight' is most probably minimal.
The contestants in this second category are listed in Chapter 14, together with
man-powered rotorcraft proposals.

## SPILLMAN AND THE LOUGHBOROUGH UNIVERSITY PROJECT

   One of the first paper proposals for a man-powered aircraft to compete in
the Kremer Competition was produced by John Spillman of the College of
Aeronautics.  Spillman presented his design study to a meeting of the Royal
Aeronautical Society in January 1962.

   The main features he advocated were more concerned with ancillary systems
than with wing and fuselage design, although component location was dictated
by aerodynamic considerations, and thus influenced the overall layout.  Spillman
suggested that the slim low fuselage, together with a reclining pilot position,
was preferable.  He also felt that the highly cambered or flapped wing sections
necessary to attain high cruise lift coefficients required a substantial load
to trim, this incurring a drag penalty of up to ten per cent.  (Wing span was
27.9 metres, wing area 26.5 $m^2$).

   Obviously also aware of Perkins' results, Spillman emphasised the need to
keep the propeller well clear of the ground, as well as trying to avoid
putting parts of the aircraft in the slipstream.  Logically a pusher propeller
mounted on a pylon which also acted as a tail fin and rudder was the outcome
of these requirements, *(see Fig. 115).*

   In view of the fact that the Kremer Competition specified a flight above
a certain altitude, Spillman proposed a take-off and cruise technique which
would facilitate the completion of the course with the minimum of peak
exertion.  He proposed that the aircraft would accelerate at ground level,
gaining full advantage of ground effect, (Spillman had used a low wing on his
design) up to a speed of about 10.5 m/sec., at which speed the aircraft would
'zoom' to a height of 3.65 metres.  On terminating this manoeuvre the speed
would drop to the cruise rate of 8.4 m/sec.

   An exercise was given to students at Loughborough University in 1964, in
which they were required to translate Spillmans' proposals into a viable design.
Their terms of reference covered design of the mechanical systems as well as the
wing and fuselage structure as a whole.

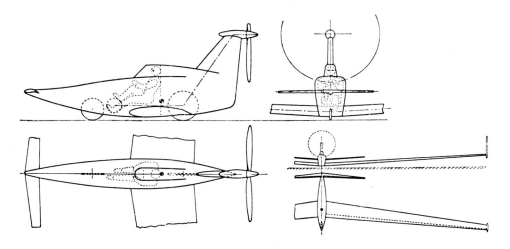

*Fig. 115* Proposed layout of the Spillman aircraft

It was found that the most effective form of wing construction was based on a birch ply D-nose with plywood ribs and leading edge skin. The trailing edge was covered with balsa wood. This arrangement was considered superior in giving a lighter weight and better aerodynamic contour than a similar wing structure covered with Melinex. The weight of this first configuration was 39.1 kg excluding adhesive weights. Fuselage strength was guaranteed using spruce, and the main spar of the fin was a combination of ply and balsa with balsa ribs. Apart from the forward part of the fin, which was to be covered with balsa, Melinex was, however, proposed for the fuselage covering.

Considerable effort was put into the design of the 'cockpit' and suspension system, the pedalling position adopted being similar to that used in the Haessler-Villinger aircraft. Both the wings and undercarriage were supported by the seat frame. It was suggested that the framework be fabricated using magnesium alloy tube welded at all joints, major members providing strength by linking the pedal axle with a bar behind the pilots' shoulders, with a second set of tubes joining the pedal wheel to the main suspension joints. A wheel similar to that used on Moulton bicycles was used as the primary ground drive, and a forward wheel was provided as it was considered that a skid system would give excessive ground friction at low forward speeds.

Initially, the designers were inclined to use magnesium alloy tubes with bevel gears for the major part of the transmission system, similar to the transmission form used in the Hatfield 'Puffin', but later proposals recommended the replacement of the bevel systems with belt drives, which could be turned through ninety degrees without reductions in efficiency.

Spillman did not carry out an independent assessment of propellers in his design study. However, using the directions given by Wickens in his comprehensive paper on propeller selection (see Chapter 15), a two bladed, 2.2 metre diameter propeller was chosen with an efficiency of 82 per cent giving a thrust of slightly over 7 kg. Loughborough suggested that the propeller be constructed

around an untwisted spruce box beam. This would provide taper, but twist would
be artificially applied by suitable shaping of the balsa ribs. Covering would
again be plastic film.

The Loughborough Group also undertook design of the control system.
Independently moving canard surfaces replaced ailerons, roll being achieved
using differential movement whilst pitch control could be effected using
simultaneous hinging of these devices. Operation of these surfaces was through
a bicycle type brake cable connected to brake levers adjacent to the hand grips
on the control column. The total range of movement was thirty degrees, and the
fact that these control surfaces were directly visible to the pilot would
probably have favourable psychological effects. Rudder movement was conventional
being obtained by moving the control column about the vertical axis, nylon cords
acted as transmitting media. A brake for ground use was provided on the main
wheel.

It was concluded that Spillman's design could be constructed with an empty
weight of 65.8 kg but much work would be required on the development of the
transmission gear and propeller before definite conclusions on the viability of
the project could be made. In 1969 the above work was reviewed at Loughborough,
but no plans for forming a man-powered aircraft group for construction purposes
were announced. This possibility was not, however, ruled out.

## THE WOODFORD AIRCRAFT

Often one comes across attempts at new endeavours, as surely man-powered
flight must still be regarded, which seem worthy of success from the point of
view of expended effort and enthusiasm of those concerned with the project,
regardless of the technical merit. If the technical features of the design are
noteworthy, as was the case with the aircraft constructed by C. Hodgess-Roper
at Woodford, then seven years of effort should be rewarded with at least an
attempt at flight. Unfortunately, this was not to be the case with the
Hodgess-Roper machine.

Preliminary data on the Hodgess-Roper man-powered aircraft was received
by the Royal Aeronautical Society in June 1961[1]. The aircraft was shown to be
a high wing monoplane, with one crew member sitting in an open frame fuselage,
from the top rear of which ran a tubular tail boom carrying rudder and tailplane. This boom also acted as a hub for the propeller, mounted just behind
the wing root. Major details are given in Table 7. As a result of discussions
with the Royal Aeronautical Society, the design was later revised. The aircraft
was given a favourable reception by several members of the Grants Committee.
No doubt this was influenced by first estimate of cost, £200, based on an eight
month timescale. However, the propeller position was attractive, and the
comparatively low wing span was thought at this stage to offer advantages in
manoeuvrability.

The Hodgess-Roper proposals were considered by the full Committee of the
Royal Aeronautical Society Man-Powered Aircraft Group at their meeting in
January 1962. No grant was recommended at this stage, further information
being required on wing profile drag and the effect that the suggested open

---

[1] *The Man-Powered Aircraft Group were aware of the project before this date, as
Rendel recorded comments on the scheme in November 1960.*

cockpit would have on the calculations of power.  The committee did, however, acknowledge that the design showed promise[1].

It is likely that the figures given by Hodgess-Roper for the cruise power were the subject of some discussion, because the maintenance of an output in excess of 0.38 kW would be only limited to little more than one minute at best. The designer did emphasise that winch launching would be used in many flights, conserving power for cruise and manoeuvres alone.

Table 7.  Early data on the Hodgess-Roper Aircraft

| | |
|---|---|
| Wing span | 18.3 metres |
| Wing area | 18.7 m$^2$ |
| Aspect ratio | 18 |
| Wing chord (root) | 1.14 metres |
| Wing chord (tip) | 0.57 metres |
| Dihedral | 0 degrees |
| Sweep | 0 degrees |
| Root twist | 2.75 degrees |
| Aileron span | 8 metres |
| Tailplane span | 2.17 metres |
| Rudder area | 0.96 m$^2$ |
| Overall length | 5.8 metres |
| Wing weight | 22.3 kg |
| Fuselage weight | 20 kg |
| Tailplane and rudder weight | 1.37 kg |
| Pilot weight | 70 kg |
| AUW | 113.7 kg |
| Maximum airspeed | 58 km/h |
| Minimum power | 0.42 kW |

Hodgess-Roper, assisted by his wife, continued working on the construction of his aircraft through 1962 and the first half of 1963, and was, under the title Woodford Man-Powered Aircraft Group, awarded a grant in July 1963

---

[1] *Prior to this meeting, Shenstone noted his conclusions on the design, which he found attractive. Although he had slight misgivings regarding the placing of the propeller, suggesting that it might be moved further aft, he found the general arrangement of interest. However, he did recommend reassessment of the weight and drag figures.*

towards the cost of construction of the machine[1].

As the project was largely a one-man effort during construction, progress was naturally fairly slow. Shenstone, reporting on the condition of the project in September 1963, stated that the wing was substantially complete internally, spars and ribs being fabricated and some components joined. The propeller blades were completed - each blade weighed 0.455 kg and the overall diameter was to be 2.28 metres. No progress had been made on the drive system, fuselage or control systems.

By this stage the overall dimensions of the aircraft had changed considerably and the pilot was enclosed in a 'pod' cockpit. The weight had increased slightly to 54 kg, and the span had grown to 24.2 metres. Corresponding wing area had almost doubled to 33 $m^2$.

Mid-1964 saw the completion of the major structural members and the transmission system, which operated using a twisted chain. The drive had variable gear ratios, obtained with various sized sprockets mounted on the main drive wheel. (Hodgess-Roper had also received assistance from Mr. Moulton, of the bicycle firm of that name).

Fifteen months later Shenstone visited Hodgess-Roper to view progress on covering the wing and tailplane, and was told that work was shortly to commence on the cockpit framework.

The assembly of the aircraft continued at a slow pace. In May 1966, at the Man-Powered Aircraft Group's sixth A.G.M., the aircraft was reportedly almost complete. However, an airfield was needed for future flight trials. At the seventh A.G.M. the report was substantially similar, although an additional drawback regarding the inability to pay the high rent necessary to obtain access to an airfield was mentioned. Aid in the form of research projects on certain aspects of the Woodford machine was being provided by third year engineering students at City University, London.

Two years later, the *London Evening News* of June 13th, 1969 carried a short article to the effect that a man-powered aircraft constructed by C. Hodgess-Roper had been partially destroyed by fire at Woodford Green. No flights had been attempted prior to this loss, but what remained of the structure was later taken to the R.A.F. College, Halton, (see Chapter 16).

Several proposals for man-powered aircraft were submitted in the early 1960s to the Royal Aeronautical Society's Man-Powered Aircraft Group for consideration. Only a few received grants, but notes on the efforts of less fortunate would-be contestants for the Kremer prize reveal several points of interest, not least the similarity of many of the designs, varying only in the arrangement of their common components, which tended also to be of similar dimensions.

The majority of the design studies showed evidence of a complete lack of economic judgement so typical of enthusiasts. Proposals for constructing and flying aircraft of a complexity comparable to the Southampton University

---

[1] *A press release issued by the R.Ae.S., dated July 25th, 1963, announced grants to the London Man-Powered Aircraft Group; the Farnborough Man-Powered Ornithopter Club (see Chapter 15); and the Woodford Man-Powered Aircraft Group. The grant to Hodgess-Roper amounted to £500.*

aircraft included cost estimates as low as £90, with timescales of the order
of three to four months.  Not all showed such naïveté, however;  one astute
businessman included a profit of £120 in his cost estimate!  Needless to say,
this was frowned upon by the Committee[1].

P.M. Savage of Kingston upon Thames put forward a proposal for a machine
similar to the original Hodgess-Roper aircraft layout.  The Committee were
critical of the wing bracing proposed - this can often cause quite a severe drag
penalty, which may easily be overlooked in initial calculations.  Wing tunnel
and structural tests were recommended at this stage.  Savage's 18.3 metre span
monoplane was a single seater.  The calculated empty weight of 27.25 kg was in
keeping with that obtained with comparable designs, and the gross weight of
91 kg provided sufficient allowance for a pilot.  A high aspect ratio of 21 was
adopted, and using a chain primary drive with belts for the secondary transmission, driving a pusher propeller located aft of the wing, the cruise power
was calculated as 0.36 kW, high for any sustained flight.  This may have been
influenced marginally by the use of an open cockpit.

One of the more unusual design studies was that completed by R.T. Wood.
This proposed a compound aircraft, with both rotor and fixed wing;  this aroused
interest but it was believed that the aerodynamic performance of the wing would
suffer as it was in the rotor slipstream.  This would inhibit the extent of
laminar flow.  It was suggested that Mr. Wood concentrate on designing an
autogiro, but the Committee were not prepared to recommend a grant to further
the work.

## THE SOUTHEND MAN-POWERED AIRCRAFT GROUP
## —ADVOCATES OF TWO CREW

The largest of the first generation of man-powered aircraft conceived and
constructed in Britain in the early 1960s was the product of the Southend Man-
Powered Aircraft Group.  This Group was formed at Southend Airport around a
nucleus of five founder members, A.G. Drescher, A.Kerry, M. Prentice,
K. Barbeary and C. Basu, who were all employees of Aviation Traders (Engineering)
Ltd., the company then responsible for conversions of large aircraft into cross-
channel air ferries for cars.

First details of a two-man machine with a tractor propeller mounted on a
pylon were put to the Committee in 1960, and were received, on the whole,
favourably.  Performance estimates were considered accurate, but Lilley
commented that the profile drag of the wing, based on the Southend assumption
of turbulent flow, was underestimated, as was wing weight.  Awareness of the
importance of a high mounted propeller was also felt in the tone of the
comments.  Shenstone felt that the proposed side-by-side seating arrangement
would increase fuselage drag unnecessarily.  He also foresaw difficulties in
perfecting the structure.  (Southend were the first design team to suggest a
tubular spar - in their case of light alloy - although this was not adopted).

Following discussions with the Royal Aeronautical Society's representatives,
a reappraisal of the design was carried out, and the design changes were
outlined in a letter to the Society in May 1961.  The main alterations were the

---

[1] *Except where stated, 'The Committee' refers to the Man-Powered Aircraft Group
Committee within the Royal Aeronautical Society.*

replacement of the tubular spar with one of conventional section, the adoption of a new wing profile, relocation of the wing which had a span of marginally under 27.5 metres, and movement of the direct drive to the ground wheel from a forward to a rear wheel.

The single wing spar was to have spruce booms with plywood ribs, located at forty per cent chord.  A NACA $65_3$ 618 laminar flow wing section was to be used, the skin being formed with model tissue shrunk and doped over balsa ribs spaced at one foot intervals.  A major proportion of the torsion strength of the wing was carried by a box nose formed by sheeting to the spar on the upper and lower surfaces with 1.59 mm balsa.  With this form of construction, the Southend design team claimed wing weights of 0.904 $kg/m^2$.

The aircraft was to have a wing area of 37.6 $m^2$ and an aspect ratio of 20, *(see Fig. 116)*.

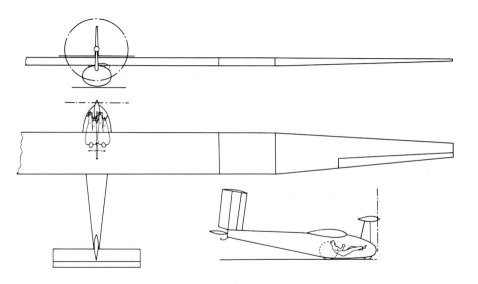

*Fig. 116*  Layout of the two-man machine
designed and constructed by a group at Southend.

Fuselage structure was to be based on a load carrying aluminium tube framework faired with balsa sheets.  The proposed transmission system employed a chain drive from a common foot crank to the rear wheel for ground acceleration. A twisted belt transferred power thence to the propeller shaft.  Conventional tailplane and rudder, the former all-moving, were proposed, and an empty weight of less than 68 kg was the aim.  A two-bladed propeller had been designed, with an eighty per cent efficiency, transmission efficiency being ninety-seven per cent.  Design rev/min was 270, crank speed being 90 rev/min.  The team calculated that two crew members, each developing 0.3 kW should enable the aircraft to attain a cruise speed of 15.25 m/sec and a rate of climb of 3.05 m/min.  The calculated turning circle was only 122 metres.

Lilley, representing the Committee, held discussions with members of the

Southend Man-Powered Aircraft Group on August 3rd, 1961, to resolve several
points arising out of the latest design submission, and to obtain an
appreciation of the Group.  The officers of the group reported that to date
funds had  been accumulated by subscriptions from members.  Facilities were
available for the construction of the aircraft in a disused bus station.  They
were hopeful of obtaining an airfield for flight trials, and completion of
manufacture of the aircraft was scheduled for May 1962.  (This date was to
prove an extremely optimistic forecast).

   The Royal Aeronautical Society favourably received the second design
submission, and on August 24th, 1961, the Man-Powered Aircraft Group Committee
informed the Southend Group that they were being given a conditional grant of
£750.  By June 1962 the central wing section (one of three 9.15 metre sections)
was structurally complete, as was the fuselage.  Units were being covered with
an aluminised plastic film.  April 1963 saw the mating of wing and fuselage,
and the fin was also in position.  No controls, cockpit or transmission
components had been fitted.  Completion date was now to be mid-1964.

   Slow progress made on the project during 1964 was attributed to machining
difficulties - it took six months to make pedal cranks and the control bar -
but it was thought that a flight could take place in July of that year.  It was
at this stage that the Southend Group became concerned about high insurance
premiums claimed by companies for the aircraft.  Eventually they obtained the
names of the Hatfield and Southampton Group brokers, and the Royal Aeronautical
Society offered £100 to cover the premium for a further year, *(see Fig. 117)*.

   Taxiing trials were completed, but it was not until July 1965 that the
first attempted flight under man-power was made, *(see Fig. 118)*.  This
attempted flight produced the following headline in the *Daily Telegraph* of
July 19th, 1965:  "Man-Powered Aircraft Fails to Take-Off:  Pedal Pin Snaps at
15 miles/h."  A brief item on the attempt followed.

   Observers from the Royal Aeronautical Society were present for the attempt,
which took place at about 4.45 a.m. on Sunday, July 18th.  The crew were
Barry Stracey, an airline pilot, and Derek Gray, a racing cyclist.  Stracey was
responsible for control while also contributing towards the power, most of
which was provided by the cyclist.  It was noticed during the first 275 metres
of the ground run that the port wing gave negative lift owing to an accidental
built-in twist, which the ailerons were unable to counteract.  Stracey noticed
that the aileron movements required excessive effort.  The newspaper report also
omitted to mention the disintegration of a tyre on the nosewheel.  This was
attributed to the very rough runway surface.

   The mechanical failure which terminated the run was caused by a crack in
a weld locating a lug carrying a crank pin.  It was felt that take-off could
take place at a speed of 27 km/h and the 24 km/h achieved on this occasion was
therefore close to that required.  Further attempts were to be made following
rectification of the faults indicated during the run.

   A. Reynolds, a member of the Southend Group who later occupied the pilot's
seat on occasions, commented after the first trial that neither pilot was at
all extended during the ground run, or was breathless afterwards.

   A modified machine had its first trials in August of the same year.  On
the third outing the chain came off and the sudden emergency stop resulted in
slight damage to the wing.  The redesigned aileron circuitry proved satisfactory

during this somewhat abbreviated test, the previous excessive loads being
reduced to more manageable proportions.  The accident occurred when the
aircraft was travelling at 24 km/h and still accelerating.  No take-off had so
far been achieved but the nosewheel had lifted, *(see Fig. 119)*.

In 1966, following further structural modifications, the aircraft was
moved to the airfield at Debden, in Essex, where it was hoped to carry out
flight trials.  By now, however, the tempo of work with the Southend Group had
slowed considerably, and it was shortly reported that the Group had broken up,
several of the members emigrating to the United States.

At the next A.G.M. of the Man-Powered Aircraft Group, confirmation was made
of the folding up of the project.  Subsequently the aircraft was stored, but
the propeller was taken over by the Hertfordshire Pedalnauts, who were commencing
construction of 'Toucan', their Kremer Competition entrant.  (This machine is
described in Chapter 16).

## LONDON — DESTINATION OF 'SUMPAC'

The disbandment of the Southampton University man-powered flight team
(Chapter 9) was not to mean the end of their aircraft.  Alan Lassière, who
worked at Southampton on this machine, had since moved to Imperial College,
London.  Obviously a lot of information and experience had been gained using
the aircraft in its original form, and in March 1963, Lassière submitted a
proposal to the Royal Aeronautical Society based on a reassessment of the design,
and the making of such modifications to the Southampton aircraft as might be
necessary.  (Lassière had been working for some time on the aircraft which he
now had at Imperial College).

Lassière, in his capacity as Chairman of the London Man-Powered Aircraft
Group, received a grant of £200 from the Royal Aeronautical Society's Group
funds[1] to enable him to proceed with modifications to the aircraft, and it was
agreed that if a successful attempt on the Kremer course was made the prize
money would be shared with the ex-members of the original team at Southampton.

The London Group was able to make comparatively fast progress, and the
first flights of the aircraft were anticipated as being in mid-1964.  Although
the pilot position was substantially identical to that of the original machine,
the framework of the cockpit was changed, in association with a new transmission
and control system.  Detail design of the latter system was considerably
improved, and attempts were made to reduce the weight of the pilot framework
by replacing the original with a riveted and welded light alloy structure.
Much of this fabrication could be done within, or close to, Imperial College,
but the major components of the aircraft were stored at Croydon.

Lassière was fortunate to obtain the services of Derek Piggott, who flew
the aircraft during trials at Southampton, as pilot for the rebuilt machine.

By July 1963 some considerable progress had been made.  The main frame and
new pylon (with less leading edge sweep) had been joined to the wing centre

---

[1] *On May 8th, 1963, the funds of the Royal Aeronautical Society Man-Powered
Aircraft Group stood at £2145, of which £583 was committed to projects.
(This excluded the Kremer Competition prize money).*

section, and the front frame of the fuselage was at an advanced state of
construction.  Instrumentation was also being re-designed.  The nose fairing
was complete, and covered with Melinex as used on 'Puffin', and a new tail unit
was in the process of assembly.  At this stage Lassière stated that the extent
of renovation and replacement of structural and ancillary components was by
necessity considerably greater than originally envisaged.  Additional design
work involving the seat, harness, wing skids and fuselage would be required.

Progress reported to the Royal Aeronautical Society in September 1963
included completion of the tailplane, and continuing work on the fin.
Renovation and repainting of the propeller was underway and the new control
column had been fitted.  However, when this was tried in conjunction with
simulated aileron loads, considerable friction in control movement was noticed.
An improved system using ball bushings for the linear movement, and ball
bearings for the aileron rotation, was being designed.  A second defect was
found when an attempt was made to mate the renovated front fuselage (based on
the original Southampton design) to the new central framework.  Complete
reconstruction of this front fuselage section had to be carried out.

By January 1964 assembly of all the major aircraft components was well
underway, but the London Group were having difficulty in finding suitable
facilities for machining several small but important detail fittings.  In an
attempt to reduce the aircraft weight, the new pylon, tailplane, fin and
front fuselage, as well as the main fuselage structure, were all covered in
Melinex.

It was not long before the financial situation of the group began to give
rise to concern.  As well as necessary expenses such as insurance, it was
hoped to carry out limited trials during the summer of 1964, and in order to
achieve this aim, it would be necessary to pay workers to carry out machining
and other minor but specialised tasks, to complete the aircraft.  It was these
arguments that Lassière used in a letter to Mr. R. Graham, Chairman of the
Man-Powered Aircraft Group Committee, requesting a further grant of £100.

At first the Royal Aeronautical Society were luke-warm towards the request,
but following the provision of more data in support of the arguments for a
grant, including a further requirement for finance to strengthen the pylon, a
sum of £120 was provisionally approved as an additional grant.

A further progress report, dated September 1st, 1964, gave a much less
encouraging forecast of the cost and timescale of future work on the project.
While major structural work was complete, Lassière estimated that six weeks of
effort was required for minor modifications and finishing, in addition to tasks
such as refurbishing the trailer and manufacturing frames for components.

Problems with the transmission system persisted;  the 'Powergrip' belt did
not run smoothly, and in order to overcome this, a narrower belt was tried, in
conjunction with guide rollers and relocated pulleys.  To some extent this
overcame the difficulty, but as jumping of the belt still occasionally
occurred, a request for money to develop a twisted 'Vee' belt system was made.

Additionally, the resignation of the London Group Secretary and the
dispersion of members as their University courses came to an end, created
recruitment problems.  It was therefore requested that support be given by the
Royal Aeronautical Society to enable a fitter to be employed full-time on
completing the assembly of the aircraft.  It was estimated that this would cost

£300, and the proposal was considered to be the only way by which the aircraft could be ready for flight during 1964.

At this stage more money was not forthcoming, and a modified request for £200 was made in March of the following year, to cover machining costs and to provide for a contingency allowance.  Graham, replying on behalf of the Royal Aeronautical Society, stated that at present the Group could not meet Lassière's request for further aid.  At the Group A.G.M. in May 1965, the progress of the London Man-Powered Aircraft Group was described as "disappointing".

Flight trials commenced late in 1965, following successful assembly and taxiing runs, but during a flight on November 12th, 1965 over the runway at West Malling airfield, in Kent, the aircraft crashed, *(see Fig. 120)*.  The pilot on this occasion was John Pratt, who attributed the forced landing to a gust of wind which sharply increased the aircraft incidence, causing it to stall. Pratt maintained pedalling power in an attempt to obtain sufficient forward speed to regain handling control, but the drop from an altitude of about 9 metres was too sudden.  The accident followed a straight flight of approximately 45 metres, and the sequence of photographs show the severe stall attitude resulting from the gust, as well as indicating the extent of damage.  The pilot was uninjured.  The final photograph was taken shortly before the machine came to rest, and subsequently the tail broke off.  Damage to the wing was severe, the main drive wheel buckled, and the metal cockpit frame was twisted.

Following this accident, enthusiasm waned and no serious attempt was made to rebuild the aircraft.

As a serious contender for the Kremer prizes, the Imperial College machine was inherently similar to, and therefore susceptible to the same flying characteristics as, SUMPAC.  This was of course due to the retention of the original wing and aileron controls, and it is unlikely that the different tail unit or pylon would have made significant improvements to the performance.

*  *  *

*Fig. 117.* The partially covered wing of the Southend
aircraft, at its conception the largest of
its type. (Span was approximately 27.5m).

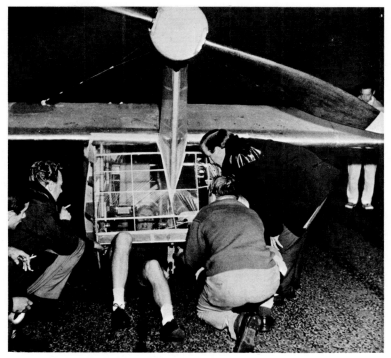

*Fig. 118.* B. Stracey, one of the pilots of the Southend
aircraft, being 'installed'. Dated July
1965.

*Fig. 119.* The Southend group aircraft following
taxiing trials.

*Fig. 120.* **Crash** sequence – the abortive flight of the
          London University group machine, 12th November
          1965.

*12*

# The
# First
# Man-Powered
# Flight
# Simulator
## — *One of Several*
## *Canadian Projects*

# *12*

The Kremer Competition, open as it was in its original form to Commonwealth entrants, was bound to attract interest on a world-wide scale, and the accompanying publicity spurred activity in North America and many European countries.  By raising the prize money initially to £10 000 and opening the competition to world-wide entry, overseas groups were given a new impetus. The introduction of a separate and easier course, with correspondingly lower prize money, restricted to Commonwealth entrants, acted as a further incentive.

## COMMONWEALTH ACTIVITY — CANADA

The first country outside Britain to show a serious interest in the Kremer Competition, and in man-powered flight in general, during this period, was Canada.  This was to be expected, bearing in mind the fact that of all countries within the Commonwealth, Britain and Canada were, at the beginning of the 1960s the only two capable of being totally self-reliant with regard to airframe and engine design and production for most aircraft types.

The Canadian equivalent to the Royal Aeronautical Society is the Canadian Aeronautics and Space Institute[1] (CASI).  It was within this organisation that a man-powered aircraft group was formed, and awareness of the interest in Canada became apparent in 1960.

A meeting was held on 10th April, 1960, in the Commonwealth Building, Ottawa, to discuss possible work on man-powered flight in Canada, with particular reference to the Kremer Competition which had just been announced in Britain. Those present suggested that universities and technical institutes might be interested in either building a complete man-powered aircraft, or supporting by research and development work a co-ordinated effort by a group formed of CAI members.  The first estimates of the cost of building such a machine were put at $10 000.

Those present were united in their criticism of the Kremer Competition rules, these being described as "not very realistic or useful".  There was general agreement that should the CASI decide to promote man-powered flight, possibly along the lines of co-ordination of sponsorship adopted by the R.Ae.S., it should not make the Kremer Competition its goal.  It was suggested that the main objective should be man-powered flight "for its own sake", but members conceded that the man-powered aircraft group(s) should be given the opportunity to enter the competition if they so wished.

With a view to promoting as much interest as possible in man-powered flight,

---

[1] *In the early 1960s the CASI was known solely as the Canadian Aeronautical Institute (CAI).*

it was agreed that the rules of the Kremer Competition should be published in
the CASI Journal, and that particulars should be sent to magazines and
organisations representing Canadian aviation.  Those attending the meeting
agreed to act as a Committee, with Dr. M.G. Whillans as Chairman.  B.S. Shenstone,
a Canadian by birth, asked to be regarded as a contributing member.

The Institute Council agreed at the A.G.M. that a committee should be
formally appointed, and on 29th June, 1960, met to define its objective:  "to
stimulate interest in, and work on, man-powered flight in Canada".  It was known
that there were pockets of interest in various parts of the country, and the
Committee hoped that it would eventually establish itself as the focal point,
using the CASI Journal as the accepted medium of exchange for information, and
as a means of arousing interest.  The Committee also considered that the man-
powered aircraft group should form a small library on the topic for members'
use, and also publish a series of papers in the Journal dealing with various
aspects of man-powered flight, to introduce the idea to members.  (Among
contributors were Enea Bossi and Helmut Haessler, who both resided in Canada
at this time).  The R.Ae.S. Group in London was kept informed of their Canadian
counterpart's activities, and occasional notes on progress were reported in the
aeronautical press in Britain.

Dr. Whillans, in his report prepared for the Committee in May 1961, stated:

"At present the Committee is of the opinion that the time is not yet right
for a full scale attempt to design and build a man-powered aircraft.  We
feel that there are many vital, but perhaps not very obvious, gaps in our
knowledge which, until they are all filled, will doom to failure attempts
to make any significant advance on what Mr. Haessler and Mr. Bossi did some
twenty-five years ago.  We believe that this country can make its greatest
contribution by a little well-chosen research and publication of results.
We must be sober and restrained in what we see to be a very formidable
engineering problem.  No other course in our opinion would be worthy of
responsible technical people.

In closing, I should like to anticipate a question which is sure to be asked.
Why is anybody interested in spending time and thought and treasure to
achieve man-powered flight, and what good will it be when they have succeeded.
I think that the only honest answer to the first question is one already
given by the Secretary when discussing the subject some months ago in the
Journal;  George Mallory's famous "because it is there", and it is not only
there;  it has been there since man first watched a bird, and it is now
tantalisingly near to attainment.  In brief, it is a challenge.

The answer to the second question - what good will it be - is not so easy
to give, but we must remember that very few people foresaw much benefit
from the Wright Brothers' achievements.  One can speculate that it might
become an absorbing sport, combining the skill of the aeronautical engineer
with that of the pilot and the prowess of the athlete.  Undoubtedly, in the
process of achieving success, a great deal of new knowledge regarding low
speed aerodynamics will be acquired.  In these days of VTOL and transition
problems, this cannot fail to be of value.

Last, but not least, man-powered flight will be quiet."

* * *

In 1960 W. Czerwinski, then an Associate Research Officer at the Canadian
National Aeronautical Establishment and an expert on glider design, who was
later to become chief designer of the major Canadian man-powered aircraft
project, introduced Canadian aeronautical engineers to contemporary thinking
on man-powered flight with an article in their Journal.

Czerwinski reviewed the work of Haessler and Bossi, before tackling the
general design features of a man-powered aircraft, In considering the structural
and aerodynamic requirements of these machines, Czerwinski assumed the following
average empty weights, depending on the number of crew members[1]:

|                |           |             |
| -------------- | --------- | ----------- |
| One seater     | 100 lb    | (45.4 kg)   |
| Two seater     | 180 lb    | (81.7 kg)   |
| Three seater   | 246 lb    | (111.8 kg)  |
| Four seater    | 301 lb    | (136.5 kg)  |

From the start, Czerwinski rejected the single seat configuration as being
inefficient and unlikely to be successful. (The definition of success in this
context does not necessarily refer to the completion of the Kremer course, but
more likely to the possibility of prolonged flights and sporting activities).
His reasons for asserting the success of a multi-seat machine were similar to
those popularly used by other exponents of the use of two or more crew, namely
the fact that one man would have difficulty in effecting control at the same
time as putting a maximum amount of effort into producing power, and secondly
that the power-to-weight ratio for a single seat machine was always less
favourable than for a multi-seater. (Czerwinski based this latter comparison
on the weight of the empty aircraft - it is more realistic to base the
comparison on the all-up weight, assuming each crew member weighs, for example,
63.6 kg (140 lb), and this does not suggest such a great advantage in adopting
a multi-seat machine. Many other factors are affected by changes in the
number of crew, and generalisations are sometimes misleading. A larger aircraft,
carrying say, four crew, would operate at a higher Reynolds number, increasing
performance. However, complexities in the transmission system could outweigh
this advantage).

In more direct contrast to the thinking on the other side of the Atlantic
at the time, Czerwinski advocated a high strength structure, designed to
satisfy a much higher load factor than that suggested by the then ARB,
detailed in Chapter 8. To implement this he proposed a D-nose wing spar, made
up with layers of fibreglass fabric arranged at 45 degrees to the longitudinal
axis of the spar. The main torsion box would be braced using a dense wood.

Aerodynamically, his proposals broke no new ground, although Czerwinski did
stress the need to develop aerofoils for the operation within the Reynolds number
range of man-powered aircraft, laying emphasis on reducing the profile drag.
Propeller design, an aspect which has received little attention in this context,
was also recommended by Czerwinski as being of considerable importance and

---

[1] *These figures compare with the weights of other aircraft as follows, (number
of seats in brackets after the name). Haessler-Villinger (1) 36.8 kg; SUMPAC
(1) 58.1 kg; Puffin I (1) 50 kg; Japanese Linnet II (1) 45 kg; Southend (2)
70.8 kg; Toucan (2) 65.8 kg; CASI Ottawa (2) 95 kg.*

worthy of much theoretical and experimental effort.

Czerwinski's concluding remarks echoed the view of the Man-Powered Flight Committee of the CASI, extending this to include more useful remarks regarding the possible path which man-powered flight might take.  He visualised such craft using their extremely low sinking speed to advantage in soaring, using weak thermals which conventional glider pilots would find insufficient for sustaining flight, although, as emphasised earlier, this would necessitate a much heavier structure capable of withstanding higher gusts etc.

## THE SMOLKOWSKI — LAVIOLETTE BIPLANE

The CASI Man Powered Flight Group Committee operated in a similar, although less ambitious way to its R.Ae.S. equivalent.  Limited funds were available, and these were appropriated on the recommendations of the Committee to give limited support to research programmes, and to support the occasional project involving the construction of a full-size aircraft.

On 17th July, 1962 the Committee met to discuss one such project, the machine designed, and by that time partly constructed, by A. Smolkowski and G. Laviolette at Calgary.  This aircraft, illustrated in *Fig. 121*, together with Smolkowski, was a single seat biplane with a three-bladed tractor propeller, driven by power supplied by pedalling action on the part of the pilot.

These two engineering graduates commenced work on the design of their aircraft in December, 1960.  The design empty weight was 38.6 kg, light alloy being used for the main fuselage structural members and foam sheeting for wing ribs and minor load and non-load carrying components.  Fuselage and wings were covered with a fabric material which required doping.  The designers were obviously greatly influenced by the standard procedure of using vertical end-plates on wings as a means of increasing the effective aspect ratio, thus reducing the induced drag.  End-plates were fixed to the propeller blades as well as on each of the biplane wings.  The Committee, in commenting on the aerodynamic design of the Smolkowski craft, rightly considered that any improvements in induced drag characteristics were likely to be more than offset by the additional profile drag of the end-plates.  Also, drag of the wing bracing would be a drawback.

Structurally, Czerwinski believed that the reduction in weight accruing to the use of foam ribs would not materialise in practice, and although the fabric was very light, doping (which was necessary because the fabric was porous) would raise the skin weight above that obtainable using proprietry plastic film.

In summing up, the Chairman stated that despite misgivings, the Committee recognised the project as a commendable effort and voted a grant of $50 from the man-powered flight fund.

The Smolkowski machine never made a free flight, but reportedly took-off towed by a car and flew for a short distance.

## RECOGNITION OF ACHIEVEMENTS

At this same committee meeting in July 1962, a scroll was presented to John Wimpenny of the Hatfield Group in recognition of the flight he had made on 2nd May, 1962.  The text was as follows:

"The Man-Powered Flight Committee of the Canadian Aeronautics and Space Institute has specified the following terms as a minimum constituting a man-powered flight:

1.  The aircraft shall be an heavier-than-air machine, deriving no lift from buoyancy.

2.  The aircraft shall be powered and controlled entirely by the crew.

3.  No stored energy shall be used for take-off or in flight.

4.  The flight shall be made over substantially level ground in substantially still air.

5.  A height of above one metre above take-off point shall be maintained for at least one minute.

Now, therefore, this is presented to J.C. Wimpenny in recognition of his flight on 2nd May, 1962, in the Hatfield Man-Powered Aircraft Club's 'Puffin', as the first man-powered flight in compliance with these terms.

Signed by the Chairman and Secretary
of the CASI Man-Powered Flight Committee"

Three months later, at Hatfield, John Wimpenny was presented with the CASI plaque by B.S. Shenstone, *(see Fig. 122)*.

## EXPO '67 — THE FLIGHT SIMULATOR

An event such as Expo '67, held that year in Canada, attracts many new ideas which, while necessarily portraying the current advanced scientific and engineering thoughts within a nation, must also prove attractive to lay members of the public.  What many will consider as an excellent and most attractive way for implementing these two aims was the unique proposal of a Canadian electronics company, Canadian Aviation Electronics, of Montreal.  This firm offered the design of a 'Man-Powered Flight Simulator', which in the words of the associated publicity literature, "will give both operators and observers a measure of the problems encountered, and a feeling of participation in the realisation of one man's oldest dreams - man-powered flight."

Canadian Aviation Electronics based their design on information received from the CASI, which was at this time well advanced with its own man-powered aircraft design.  The simulator model was slightly modified, and in order to stand up to the wear and tear to which the unit would obviously be subjected, was designed to be extremely robust.

Although the simulator was never built, the publicity brochure written during its design described most vividly the experiences of the operator and spectator:

"The approaching visitor will see a small aircraft of rather novel design, poised for take-off on a platform.

If a flight is in progress, a passing visitor will see the aircraft apparently floating in the air with its propeller turning. In front of it on a visual screen, he will see a replica of the view seen by the pilot of an actual aircraft.

Visitors who wish to operate the simulator will enter the cockpit ... and sit on a structure similar to a tandem bicycle. They will see through the windscreen a picture of the terrain, runway, etc. and on a panel below the windscreen simulated altitude and airspeed. As they operate the pedals, the scene will alter. When sufficient speed has been developed, the view will be further modified in perspective to simulate gain in altitude. As the aircraft altitude increases, the simulator will move vertically through a total travel of approximately two feet (0.61 metres). This will give a 'live feel' to the operators within the simulator. If they cease to pedal sufficiently rapidly, the simulator will slowly descend. Depending on the work done by the operators, it will be possible to fly a complete figure-of-eight track around two pylons, as required for the Kremer Competition."

The accompanying technical specification of the simulator stated that ordinary bicycle handle bars would be used, but these would not operate fully as a control column. Effort was to be the sole criterion when judging the duration of flight, and a combination of effort and manoeuvrability was not catered for. Also, no doubt to placate the 'environment lobby' the proposal stated that "no sound track will be proved, as the operation of a man-powered aircraft is essentially silent". Perhaps a recording of the pilot puffing and panting with his exertions would have been more appropriate!

## THE OTTAWA MAN-POWERED FLIGHT GROUP

The most important project in Canada is that of the Ottawa Man-Powered Flight Group, under the design leadership of Czerwinski. Established as the Ottawa Group of the CASI Man-Powered Flight Section (this was formed out of the man-powered flight committee set up within the then CAI in 1960), shortly after the formation of this latter unit, the group was initially concerned with the specification of likely areas in the field of man-powered flight where specialist knowledge could be applied, for example as university research projects. Among features investigated in such a way were the power outputs obtained in the cycling position, and structural design, particularly of the wing.

The latter aspect was dealt with by Czerwinski, who was thus given the opportunity to expand his earlier studies on this topic. In 1966 he wrote a paper on his deliberations which was published the following year in the R.Ae.S. Journal. Introducing the problems and potentials of successful structural design of man-powered aircraft components, he rightly states that as the type of structure with which one is dealing is so removed from conventional powered machines, or even gliders, experience gathered has been limited and the potential for improvement should be abundant.

Before making positive suggestions, Czerwinski recommends the rejection of stressed skin structures as a solution to the problems peculiar

to a man-powered aircraft framework, stating that such assemblies are inefficient
in cases where working stresses are very low. He advocated[1] the use of tubes
as being particularly attractive, but considered that, as in all structures
fabricated from simple units, there was always the problem of joining them
efficiently into a framework. Typically, one would use gussets bolted or
riveted to each member which was to be joined at one point. Welding would be
an alternative where the materials were suitable. Czerwinski emphasised that
the main drawback of these types of joints was the reduction in strength caused
by the drilling of holes in the tubular members for fixing the gussets, or by
welding of the components, where heating may change the properties of the
material. Another disadvantage of these jointing techniques, particularly
welding, was that they could not easily be used in conjunction with very thin
gauge materials of the type likely to prove sufficient for application in a
man-powered aircraft.

In conjunction with his work in the field of very low density structures,
Czerwinski succeeded in developing a joint which he claimed would allow strong
and rigid  junctions between thin walled tubular sections to be fabricated.
He proposed to join high strength aluminium alloy tubes with wall thicknesses
of the order of 0.4 mm, these being too fragile to drill or weld and retain
strength, using a resin-impregnated fibreglass cord. The cord was wrapped
around the joint, the tubes having previously been shaped to fit neatly together
in such a way that tensile, shear and torsion loads were satisfactorily borne
by the finished joint. Joints involving six members were manufactured using
this technique, and Czerwinski recommended its use to amateur builders, as he
believed it to be simple to perfect as well as requiring the minimum of tooling,
(see Fig. 123).

Czerwinski noted two other potential developments in the structure of man-
powered aircraft which could possibly reduce weight. One, the D-nose wing spar
which he earlier advocated, was a typical element subjected to bending and
torsion, and it was suggested that a good structural combination could be
achieved by using the high strength material for taking bending loads as spar
caps, and a low density material for shear webs. A D-nose spar designed by
Czerwinski along these lines was shown to exhibit significant weight savings,
mainly due to a 25 to 30 per cent reduction in the weight of the spar caps by
using high strength aluminium alloy tubing instead of spruce.

Many of the structural innovations proposed by Czerwinski were incorporated
in the design of the Ottawa man-powered aircraft, which has been slowly
progressing through design, development and construction stages over the past
decade.

The Ottawa aircraft, a general arrangement drawing of which is shown in
Fig. 124, is unique when compared with other contemporary serious designs in
possessing two propellers mounted in the wing, although it resembles the Bossi-
Bonomi machine in this respect, the latter having employed two tractor
propellers.

---

[1] *Czerwinski was writing his paper in the light of the flights of 'Puffin II',
which was an excellent contemporary example of near-optimum use of light-
weight structural materials.*

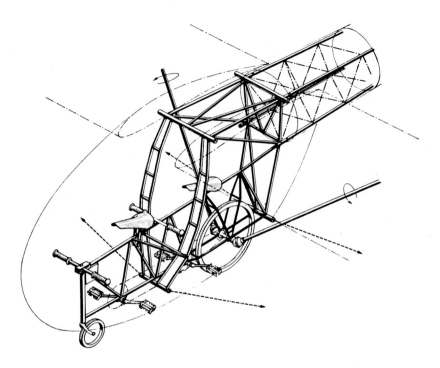

*Fig. 123*  The main frame and power train of the Ottawa man-powered aircraft.

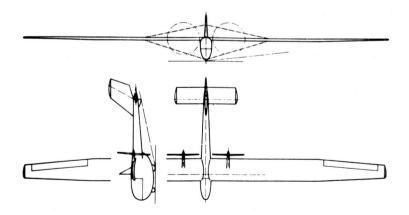

*Fig. 124*  An arrangement drawing of the Ottawa man-powered aircraft 'currently under construction'.

In choosing a two man crew[1], the Ottawa design team were partly influenced by experience with 'Puffin', which suggested that during trials, particularly at the very early flight testing stages, the pilot had little time or inclination to concentrate on controlled flying of the machine, being preoccupied with producing sufficient power to remain airborne at a reasonable altitude. A second feature which was a pointer towards having a two-man crew was that the minimum cruise power occurs over only a very narrow speed range, and this speed should be maintained precisely during the flight. If deviation from the speed occurred, either increasing or slowing the aircraft velocity, the crew would be required to exert more power. One crew member alone would possibly not be able to spare enough concentration to monitor the speed and adjust where necessary. These factors led Czerwinski to choose an expert pilot with a good average cycling ability, and a professional cyclist who could provide propulsion power well in excess of average, as being the model combination.

The Ottawa Group felt that in addition to the advantages claimed above, regarding selection of two crew, there were indications that a two-seat machine would weigh less per crew member than a single seater[2]. They felt that the cockpit structure would be more efficient and that aerodynamic efficiency would also be proportionally better, as the crew could be arranged in tandem, maintaining an identical frontal area to that of a single seat aircraft. (The wetted surface of the fuselage would increase, however).

Table 8. Main dimensions and weights of the Ottawa Aircraft

| | |
|---|---|
| Wing span | 27.4 metres |
| Wing area | 41.8 m$^2$ |
| Aspect ratio | 18 |
| Tailplane area | 5.3 m$^2$ |
| Fin area | 2.79 m$^2$ |
| Wing weight | 41.4 kg |
| Fuselage weight | 4.77 kg |
| Propeller and transmission weight | 9.54 kg |
| Total empty weight | 74.9 kg |
| Crew weight | 129.5 kg |
| AUW | 204.4 kg |

[1] *About this time an American experimenter wrote to the CASI suggesting a three-man aircraft. The pilot would sit in the fuselage, and 'pods' in each wing would contain the two other crew. Their combined efforts would be used to drive a single tail-mounted propeller. (See also Chapter 16, the 'Newbury Manflier').*

[2] *Czerwinski gives figures to back this argument, and these are shown earlier in this chapter.*

The propulsion system, mentioned above, consists of two contra-rotating pusher propellers mounted on the trailing edge of the wing. It was felt that the adoption of two propellers would allow a very short mechanical transmission path to be used, and would also take out any asymmetry which could affect handling. In practice this latter characteristic has not been predominant in single-propeller machines. A standard racing bicycle pedal crank and chain combination, a set being supplied for each crew member, is used to drive the rear wheel axle, as on a conventional bicycle. Extension shafts projecting from each end of the axle contain angular gear boxes from which lead the drive shafts to the propellers. Further gear boxes are located in each propeller hub, for the second direction change, each gearbox being essentially a universal joint. As a protection against sudden changes in speed caused by the aircraft coming into contact with the ground and imparting additional energy to the transmission, a ratchet arrangement is incorporated to provide an override. A post-graduate engineering student at Carleton University undertook a practical test determining the angle drive transmission efficiency. An average figure of 96 per cent was found.

Much of the front fuselage section encapsulating the crew was based on a structure built up using aluminium alloy tubes joined by Czerwinski's fibreglass cord/epoxy resin combination. Even the joints between the pedal crank bearing case and the four tubes extending from this component to the front wheel, handlebars, seat and rear wheel respectively, were fabricated using this technique.

The choice of wing section for the Ottawa aircraft was apparently difficult because the group were unable to trace data on low drag profiles exhibiting satisfactory characteristics at the low Reynolds numbers of man-powered aircraft. It was therefore proposed that the group design their own section, and this was implemented, the design procedure being checked by wind-tunnel tests at a Reynolds number of $0.9 \times 10^6$. Characteristics of this aerofoil are given in Table 9.

Table 9. Original Ottawa aerofoil data

| | |
|---|---|
| Reynolds number | $0.9 \times 10^6$ |
| Maximum $C_L$ | 1.65 |
| Minimum $C_D$ (at $C_L = 1$) | 0.011 |
| Thickness/chord ratio | 0.12 |

Later, however, the Hatfield Man-Powered Aircraft Group disclosed to the Ottawa designers technical information on the Wortmann aerofoil section used on 'Puffin II'. This had a lower profile drag and greater thickness (13.6 per cent) than their own section, and is being used instead of the latter. The extra wing thickness was particularly attractive because a stronger torsion box, in the form of a D-nose, could be made for an equivalent, or possibly lower, weight. Wing bracing, in the form of two profiled high tensile steel tapes, is used; this normally is rejected because of the drag penalty, but the Ottawa group have calculated that the power penalty of additional drag in this case will be more than countered by the weight saving in the main spar.

By mid-1966 design work on the Ottawa machine was well advanced, and detail design of the transmission and front portion of the fuselage had been completed. Several components were being manufactured, including the angular gearboxes, which are among the most expensive items.  In completing design of the fuselage and transmission, the group were aiming at early manufacture of these systems so that the cockpit and propulsion system, fitted to wing stubs, could be used as a ground rig for training, efficiency measurements and structural tests.

The late 1960s saw little progress on the project, mainly due to financial difficulties and loss of work-space.  Under the Chairmanship of A.N. le Cheminant work in 1970 and 1971 gained some momentum.  This was in spite of the reorganisation necessary owing to the disbandment of the CASI Man-Powered Flight Section in January, 1970.

Propeller design undertaken by Czerwinski was completed in 1969 and templates for the construction were recently being manufactured.  The assistance of the Weybridge BAC Group was enlisted in obtaining a manufacturer of the ground wheel, in the form of a racing bicycle wheel.  The financial position has improved, but the Ottawa Group is unable to give estimates as to the dates of 'roll-out' and first flight trials.  As the design remains in essence identical to that first conceived at the beginning of the 1960s, the group's chances of meeting the performance requirements of a flight equivalent to the Kremer course appear to be diminishing.  Their main hope must lie with the possible advantages claimed for the two-seat configuration, so far largely untried over any significant distances.

* * *

*Fig. 121.*   The Smolkowski biplane, one of the few man-powered aircraft constructed in North America.

*Fig. 122.*   Presentation of the CASI plaque to J.C. Wimpenny at Hatfield on 21st August 1962. Mr. B.S. Shenstone, facing Mr. Wimpenny, made the presentation in his capacity as a corresponding member of the CASI Man-Powered Aircraft Committee.

# 13

# Projects
# in
# Japan,
# Austria,
# South Africa
# and the
# United States

# *13*

The interest in man-powered flight within Canada has been most akin to that fostered in Britain by MAPAC and the R.Ae.S. in the late 1950s and early 1960s. Official aeronautical engineering bodies gave moral and practical support to groups and set up guidelines as to the basic requirements of a man-powered aircraft. Good technical support, albeit in a fairly minor key was provided by universities and colleges, and projects became considerable team efforts involving sub-contracted research work and production. The Ottawa Group is fully representative of this development, which has, however, progressed at a faster pace within Britain.

Publicity associated with the setting up of the Royal Aeronautical Society Group, and the Kremer prize money, provided an incentive overseas, but not all interested in the subject reacted in the same way. Some groups constructed machines with the Kremer course as the sole objective, others recognised the difficulties in attaining such an ambitious goal, and settled for more modest aims, with flights of a hundred metres or so.

Interest was, and remains, world-wide, ranging from Japan to Austria, and the United States of America to South Africa.

## SOUTH AFRICA — VINE

It was in South Africa that almost immediate interest in the Kremer Competition was visible. The influences of the political difficulties at the time were evident from the point raised in a letter received by the R.Ae.S. in September, 1961, from S.W. Vine of Krugersdorp. Mr. Vine was worried about his eligibility for the Kremer Competition, (at that time open only to Commonwealth participants), because during the period of construction of his aircraft, South Africa inconveniently opted out of the Commonwealth. As he was shortly to attempt a flight, Vine queried his legal standing, but there was no note as to whether this was resolved. The outcome would have been of academic interest only as Vine's machine was not wholly successful.

Vine was born in London and was appointed to Lord Gladstone's staff when the latter was Governor General. His interest in flying began at an early age, and he was designing and building gliders in the late 1920s. He achieved a soaring flight of six and a half hours in one of his gliders, the first such flight in South Africa. In 1932 Vine graduated to powered aircraft and built three in the following two years; all had 30 kW (40 hp) Scorpion engines.

Following retirement from production engineering in 1961 he embarked upon a seemingly quite hectic design and building programme involving three man-powered aircraft. The first of these machines was intended as an experimental unit with a very limited capability, and it proved to be just this, resembling a high wing monoplane with lifting rotors mounted in each wing. Vine's second prototype *(Fig. 125)*, was constructed along more conventional lines, with a parasol wing and nose-mounted tractor propeller. The main features of this aircraft are given in Table 10. It is interesting to note that the wing section chosen, a modified Göttingen 535, was identical to that on the Haessler-Villinger aircraft.

The aircraft was flown for the first and last time on 17th May, 1962, and a flight of 183 metres was claimed. However unaided flight with such a small wing area would be marginal. In describing the flight, Vine recalls a take-off run of about 140 metres, followed by a period during which he was airborne, the aircraft pointing into the wind. After flying an estimated 180 metres, a sudden gust lifted the aircraft several metres and created a stall altitude from which the pilot was unable to recover. Vine escaped with cuts and bruises, but the aircraft was written off, having nose-dived into the ground. (Some accounts of the events of 17th May, 1962, suggest that the aircraft suffered its accident on its third flight, two previous flights on the same day having been more successful. The flights were made at Randfontein).

Table 10. Main dimensions and weights
of the S.W. Vine aircraft

| | |
|---|---|
| Wing span | 12.2 metres |
| Wing area | 20.4 $m^2$ |
| Wing section | Gö 535 (modified) |
| Aspect ratio | 8 |
| Dihedral | 6 degrees |
| Taper ratio | 1:1 |
| Tail unit | Conventional |
| Fuselage length | 5.5 metres |
| Take-off speed | 37 km/h |
| Wing weight | 38.6 kg |
| Fuselage weight | 54.5 kg |
| AUW | 170 kg |

In a letter to London dated 31st July, 1963, Vine confesses that he was unwise to attempt a flight in the rather windy conditions, but as South Africa was due to leave the Commonwealth on 31st May, 1962, he wanted to make certain that an attempt on the Kremer prize was made before his eligibility came into question.

At the time of his flight, Vine was over seventy years old. Construction had taken nine months, and much of the time had been spent on perfecting propeller and transmission design, Vine aiming for a very high propeller rotational velocity. Hand cranks were used in conjunction with pedals for producing power.

Vine proposed to construct a third man-powered aircraft, with a pusher-propeller on the tail and a different wing section, but this was never completed.

* * *

Even if it has not produced successful man-powered aircraft, the Southern Region of Africa certainly generates conditions conducive to high work output. For example, a thirty-nine year old engineer and powered-aircraft pilot made no less than twenty-one test rigs to check the validity of his man-powered aircraft system and component design. Philip Badisch of Turffontein constructed his aircraft in mid-1966 in an attempt to win the still-elusive R.2000 offered by L. Franklin[1], a Johannesburg businessman, for completing what in effect is a copy of the Kremer course.

Badisch, another exponent of high propeller speed, this time aiming for 1300 rev/min, made his aircraft, which was a fixed wing monoplane, out of aluminium and fabric. Pedalling was used to provide drive for the propeller, and take-off was anticipated at a speed of 43 km/h.

## RHODESIA — THE 'LIMBA' PRODUCTION LINE

Two potential fliers are reported from Rhodesia. Gina Ruggero Sachi constructed a 'Helibike' with four rotor blades, each having a length of 3.35 metres and 0.91 metre chord. Materials used were bamboo and silk. Flights of three to five minutes were claimed, but no reliable accounts support such an achievement.

A Rhodesian schoolmsster, Sanderson Chirambo, constructed several machines in his 'Limba' (flight) series, commencing in 1963 when he read of the Kremer and Franklin prizes. 'Limbas I' to 'IV' had fixed wings, and all failed to fly. 'Limba V', an autogiro assembled from a collection of bicycle parts, also remained on the ground. A sixth prototype, similarly of autogiro configuration, was proposed but lack of funds terminated work. Chirambo contemplated moving to Malawi in the hope of receiving government financial support, which was not forthcoming in Salisbury. (Government assistance there did extend to a flight in a helicopter and a book on aircraft construction!)

## AUSTRALASIA AND THE PASSENGERS WHO 'WORK THEIR PASSAGE'

The work on man-powered flight in Australia and New Zealand, like that in South Africa, has been largely unco-ordinated. This does not encourage the formation of groups of experts to design and construct an aircraft, and it is generally left to enthusiastic amateurs to pursue an aim such as the winning of the Kremer prize.

One of the first contacts made by an Australian with the Man-Powered Aircraft Group of the R.Ae.S. was in the form of a letter dated 21st September, 1960, to B.S. Shenstone. The correspondent, P. Van Hessen of Sydney, proposed a balloon-assisted man-powered aircraft. He claimed that with the aid of balloons the weight of the aircraft could be reduced by approximately 50 per

---

[1] *A short article in the 'Johannesburg Star' of 4th August 1960, referred to Franklin, a former British champion miler, and his possible flight from the roof of a building in the city, aided by aluminium wings. His job at the time was an underwriter for a life insurance company!*

cent.  Van Hessen allowed his imagination free rein when describing the
possible applications and layout of such a machine.  Multi-seat aircraft with
ten to twenty crew, cranking in addition to pedalling, would provide power to
sustain a high cruising speed.  The extra safety factor created by the
presence of the balloon would be welcome, and it was also suggested that a hook
be provided for a mid-air catch by tow-plane.  With one eye on the commercial
possibilities, Van Hessen preferred an aircraft which would carry a number of
passengers, although it is not certain whether they were expected to 'work
their passage' in part payment!  Applications proposed included gliding, sight-
seeing, bird-watching, weather study and exercise!

A project which went further than the paper proposals of Van Hessen was the
aircraft constructed by Barry Ranford of Riverton.  This was a conventional
machine with an all-up weight of 158.5 kg, rather high when it is considered
that the wing span was only 9.1 metres.  The fuselage was a metal and wood
frame covered with fabric, and had an overall length of 6.4 metres.  No flights
were made.  The *Sunday Times* of 14th January, 1962, reported a project in Perth,
Western Australia, the aircraft being a high wing monoplane with a nose-mounted
tractor propeller.  The machine was characterised by end plates of approximately
0.3 metres depth at each wing tip.

Patricia Bonham and Hermes Celio, natives of Sydney, completed construction
of a small helicopter which would utilise muscle power.  The helicopter,
designed for an attempt on the Kremer course, and named the 'Spinaloft', was
fabricated using plywood.  It weighed slightly over 45 kg.  The pilot, Celio,
who was once a French glider champion and later Chief Engineering Instructor
and test pilot for the Egyptian Royal Aeronautical Federation, operated the
rotor via a system of treadles.  A form of jet propulsion was proposed however,
the treadles being used to drive a compressor, the air from which was passed
through the rotor blades, each 3 metres long, and expelled at the tips.  Pitch
adjustment was catered for, and sufficient lift was claimed to be generated
when the rotational speed reached 200 rev/min.  The theory was, as in so many
similar cases, not borne out by the experiment.

The only man-powered aircraft in the Southern Hemisphere to sport a two-
man crew is believed to be the machine constructed by F. Lindsley and D. Little
of Auckland, New Zealand.  With a span of only 15.2 metres and a wing area of
46 m$^2$, the craft has a very low aspect ratio.  The empty weight of 91 kg is
also high for an aircraft of such small size, although plastic foam and balsa
wood were reportedly used widely in the structure, and a plastic film was used
for covering.  No flights of note were made.

## THE UPENIEKS ORNITHOPTER

There have been several serious studies in Australia of the ornithopter as
a means of achieving man-powered flight.  Upenieks[1], working in Perth, has
constructed an oscillating wing machine which, while it failed to fly due to
structural failure, indicates a more sophisticated approach to this form of
flight than that adopted by some others in this field.

Upenieks' philosophy was based on three main aims:  to reduce friction in

---

[1] *H. Upenieks.  Man-powered flight:  the oscillating wing machine.  The
Aeronautical Journal, Vol. 77, No. 754, October 1973.*

those mechanical parts of the power train operating at high speed;  to eliminate
what he claimed to be the disadvantages of a conventional propeller;  and to
free the crew from the task of activity at or near the point of physical
exhaustion without opportunity to rest or recuperate during the exercise.  He
also believed that flapping wings could significantly reduce the complexity of
the aircraft, widening the scope of man-powered flight as an activity to be
enjoyed, possibly as a sport, by many.

   The most important feature of the oscillating wing machine proposed by
Upenieks is the use of a balanced power system.  This balanced power system,
which may be crudely likened to a flying pogo stick, has three main components
- the all-up weight, the elastic support assembly and the weight and muscular
power of the crew member.  Its incorporation in a practical assembly is
illustrated in *Fig. 126*.  Upenieks makes a distinction between the state of
'no lift', when the machine is on the ground, as in *Fig. 126a*, and the state of
'lift' when it is in flight, *(Fig. 126b)*.

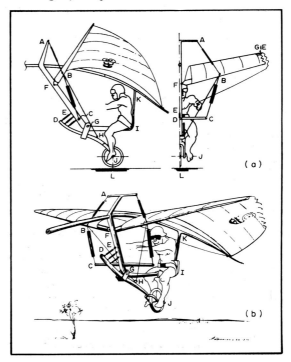

*Fig. 126*  The oscillating wing system developed by Upenieks.
This shows (a) the balanced power system on the ground, and
(b) in flight.

   When the machine is on the ground, the weight is balanced against the
elastic component DE when the centre of gravity is directly above point L, and
the crew weight is supported at point I in *Fig. 126a*.  The application of the
pilot's weight at point I and point J in an alternating rhythmic manner will
cause the aircraft to rise and fall.  In flight the all-up weight which acts

at the centres of effort (CE) in the wings is balanced against the elastic
components BC and DE shown in *Fig. 126b*.  The power potential of the DE and
IJ components is exploited via BH.  The system is in balance when the wings
come to rest at the desired dihedral angle, and it is no longer necessary for
the pilot to apply physical effort to keep the wings in this position - in
other words the aircraft becomes a glider.

It is generally acknowledged that the most difficult problem in ornithopters
or oscillating wing machines is the correct simulation of wing motion and the
associated construction methods to incorporate this motion.  The work of
Betteridge and Archer, briefly described later in this chapter, and the
drawings prepared by D.V. Curry (presented in Chapter 16) emphasise this
difficulty.

The wing form chosen by Upenieks is shown in *Fig. 127*.  It is divided into
three segments, each of which has a different function.  Area 1, the lift area,

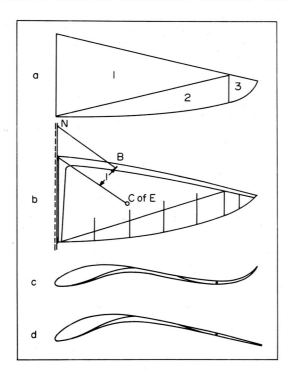

*Fig. 127* Upenieks' wing, showing the plan form and the
anticipated movement of the trailing edge during flight.

provides structural strength and houses the elastic components AB and BC.  The
power area, 2, provides the means of power utilisation for propulsion by
flexing of the trailing edge to the degree shown in the lower sketches, and
alteration of the angle of attack.  The wing control area, 3, provides the
means to give the aircraft lateral stability.

Upenieks foresees a simple membrane-type structure providing the majority of the wing surface, with local stiffening at the leading edge. While this appears to have succumbed to structural failure in the prototype, and strength is a major problem in any ornithopter-type structure, Upenieks argues that hang-glider type construction, with the additional leading edge bracing, could suffice. Certainly it would be very attractive to man-powered flight devotees if a simple development of a hang glider, say with oscillating wings, could be successfully operated. However the experience of all but the 'conventional' propeller-driven man-powered aircraft suggests that gliding in any form, from both the aerodynamic and structural points of view, is further removed from man-powered flight than commonly believed.

Although of a much more theoretical nature, the study by Betteridge and Archer[1] also reveals current thought on the possible application of ornithopters as man-powered aircraft. Full treatment of their work is not within the scope of this book, but a number of their conclusions will be of interest to the devotees of this form of flight. They found that a lift distribution could be obtained on a flapping wing which would produce thrust and, perhaps of greater significance, this lift would remain constant throughout the flapping cycle. Betteridge and Archer also found that, under certain conditions, a man-powered aircraft using flapping wings would have a higher propulsive efficiency than the more conventional propeller-driven type, with propulsive efficiencies in excess of 85 per cent being obtainable.

Unfortunately the analysis is restricted to a rigid wing, not taking into account the varied movements observed during bird flight. The extension from theory to practice is therefore unlikely to benefit man-powered flight development although the theory itself relies on well-tried aerodynamic bases.

## POST-WAR GERMAN ACTIVITIES

The post-war German endeavours in the field of man-powered flight are to date insignificant when compared with the work done in the 1920s and 1930s. The principal 'émigré' exponents of man-powered flight, Lippisch and Haessler, are now resident in the United States and Canada respectively, where conditions in the 1950s allowed them to pick up relatively quickly the themes on which they were concentrating in the 1930s. Ursinus' Muskelflug Institut had disappeared, and the Haessler Villinger 'Mufli' was destroyed during the war. As serious attempts at designing man-powered aircraft are best carried out by those well-versed in at least some aspects of aeronautical engineering, one would not expect any important work to emerge until a viable aircraft industry had been established for some considerable time. Thus perhaps during the next decade we will see innovation in the design of man-powered aircraft within Germany[2].

Only two or three post-war German proposals for man-powered aircraft have come to the attention of the Royal Aeronautical Society in London. The earliest reference was made in May, 1955, in the *London Evening News*, concerning a

[1] *D.S. Betteridge and R.D. Archer. A study of the mechanics of flapping wings. The Aeronautical Quarterly, pp 129 - 142, May 1974.*

[2] *Lippisch recently designed a small man-powered monoplane which ill-health prevented him from constructing. This is described later in the chapter.*

certain Herr Budig, a mechanical engineer who lived in Wallhausen and had designed a glider with which sustained flights could be made by flapping the outer parts of the wing.  The aircraft apparently progressed no further than the drawing board, as did the machine designed by William L. Fischer of Munich.

Fischer wrote to the R.Ae.S. Man-Powered Aircraft Group in September 1963, requesting that they comment on his proposals for a biplane.  It had been the intention of Fischer to concentrate on an ornithopter, believing this to be the most promising solution available, but his scheme was rearranged owing to the fact that he was unable to obtain support within the aircraft industry for such an 'unconventional' system.  Fischer also felt that the oarsman position was most conducive to high power generation, but he adopted the cycling technique as the basis of his design study, admitting that the successes of the Hatfield and Southampton aircraft had influenced his choice.  His design consisted of a streamlined pod fuselage in which were two cyclists mounted in tandem, the fuselage being supported on a tricycle undercarriage, one wheel of which was driven.  A 2.44 metre diameter propeller was attached to the back of the fuselage pod, driven by belt to a bevel gear, (primary drive was a chain). Above the fuselage was a biplane structure based on a Chanute glider, and a conventional tail assembly was joined by a flimsy boom to the wing trailing edges.  Fischer chose a biplane configuration because he believed it to be lighter than an equivalent monoplane, in effect a biplane wing being a box section if sufficiently well-chosen bracing is used.  In addition, it offered the advantage of low wing-span.  He was aware of the potential drag penalties of bracing wires and struts, but calculated that these would be insignificant at the 40 km/h cruise speed of his machine.

*  *  *

One of the most sophisticated and successful man-powered aircraft in Europe to date is that constructed by Josef Malliga in Austria in 1967.  In configuration a low wing monoplane with a twin tail boom, the Malliga aircraft had a span of 19.8 metres and an empty weight of 51.2 kg.  Apparently designed with comparatively short flights in mind, as opposed to the Kremer course requirements, this aircraft flew a distance of 200 metres at an altitude of approximately one metre during the Autumn of 1967, with Seigfried Puch, a gliding instructor, at the controls.  Since that date distances of the order of 350 metres have been flown with unaided take-off, *(see Figs. 128 and 129).*

## THREE RECENT ITALIAN DESIGNS

Parallel developments were to be anticipated in Italy,  Enea Bossi had emigrated to the United States and his aircraft was destroyed.  With complete changes in the order of priority regarding developments in the 1940s and 1950s, it was not until 1962 that press reports of activity in the design and construction of man-powered aircraft in Italy were noted.  One such report recalled Umberto Carnevali's attempt to fly in March 1963 at Urbe Airport near Rome.  The outcome of the tests with his ornithopter may be deduced from the fact that he was later made an honorary non-flying member of the local Aero Club!

Alberto Lavazza of Genoa exchanged correspondence with the R.Ae.S. concerning his design study for a fixed wing machine and his theoretical studies on flapping flight.  Although he suggested that the Society could usefully exchange information with him, the offer was not taken up, and no reports were received regarding eventual construction of the Lavazza aircraft.

Considerable inventiveness was exhibited by Robbiati Cesari in his work on man-powered aircraft design during the period 1962 to 1964. Cesari made a model of his monoplane which had a pusher propeller mounted at the tail, the latter being of cruciform shape. In order to take full advantage of the aerodynamic performance obtained by locating the propeller at the tail, the pilot was in an almost fully prone position, lying on his back, with his head on a bolster in the nose portion. Thus he could not see directly ahead. This enabled a very thin cross-section streamlined fuselage to be used, in conjunction with short and simple transmission runs. To preserve simplicity, no driven ground wheel was provided.

Table 11. Principal dimensions of the Cesari aircraft.

| | |
|---|---|
| Wing span | 16 metres |
| Root chord | 0.855 metres |
| Root section | NACA $65_4$ 421 |
| Tip chord | 0.49 metres |
| Tip section | NACA $65_4$ 021 |
| Wing area | 11 $m^2$ |
| Fin area | 0.9 $m^2$ |
| Tailplane area | 1.5 $m^2$ |
| Tailplane span | 3 metres |
| Wing weight | 16.5 kg |
| Fuselage weight | 14.5 kg* |
| Tailplane weight | 0.4 kg |
| Empty weight | 31.4 kg |

* includes tail fin and propeller weight

A form of pedalling action was used to generate power, but oscillatory rather than rotational motion was imparted to chains which led to a shaft on which was mounted a bevel gear. The propeller shaft ran on three ball races and a bronze bushing. The wing structure was based on a single spar made using sheets of obeche wood, the core consisting of 'Cellovel' strengthened by canvas and saturated with 'Aerolite'. 'Cellovel' wing ribs were used, and the whole wing surface was covered with the same material. Obeche was used for support at the leading edge and balsa served this purpose at the trailing edge. An obeche framework formed the basis of the fuselage. Controls were conventional, but the ailerons were noticeably short, and would probably have proved inadequate, *(see Fig. 130)*.

## THE INFLUENCE OF LIPPISCH IN THE UNITED STATES

Man-powered flight has never attracted a strong following of serious aeronautical engineers in the United States, and even the inception of the International Kremer Competition did not result in any noticeable enthusiasm, even within the aircraft industry.

Although the aircraft designed in America by Lippisch was never constructed,

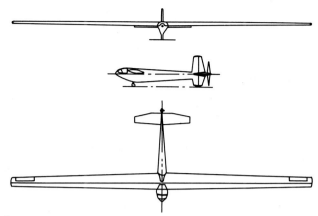

*Fig. 130* Arrangement drawing of the Cesari proposal
which only reached the model stage.

he did make his mark in circles close to the subject as far as the philosophy
behind attempts in the decade following the Kremer prize was concerned.  His
views were echoed by other serious workers in the field: "Man-powered flight
would create a new type of sport, in the same vein as other Olympic sports.
It would create an air sport, and the aircraft to accomplish this would be
sports implements, as is the racing shell built for rowing competitions".
Lippisch felt that the Kremer course was too ambitious a goal with the
technology available.

His own proposal was for a single seat machine made in the main of balsa
wood.  Some construction work was reportedly carried out on this aircraft, but
the state reached is uncertain.  The aircraft was novel in that the complete
wing was to be rotated about the spar to effect roll and pitch control.
Lippisch considered that a chain drive was the most efficient, and a long
twisted chain was proposed to drive the pusher propeller located at the back
of the pod fuselage.  The pilot was to be in a semi-reclining position beneath
the high all-moving wing, and the boom supporting the orthodox tail arrangement
ran from the bottom of the fuselage.

Table 12.  Main dimensions of Lippisch design

| | |
|---|---|
| Wing span | 15.2 metres |
| Wing area | 19.4 m$^2$ |
| Wing section | Göttingen 682 |
| Aspect ratio | 12 |
| Fuselage length | 5.5 metres |
| Empty weight | 23–27 kg |
| Propeller diameter | 2.0 metres |
| Propeller speed | 200–250 rev/min |

Before concentrating on a single seat aircraft, Lippisch investigated the
possible advantages of a multi-seat arrangement. He estimated that the weight
of the structure would increase by only 50 per cent and thus the power-to-weight
ratio must be more favourable, particularly if the second crew member could
assist with his arms. Such an aircraft would have a canard wing arrangement
with a pusher propeller at the tail, and would be inherently stable.

Aerodynamically, Lippisch felt that an all-wing machine would be the most
suitable. He claimed that instabilities which were once considered inhibiting
could be overcome, based on experience gained with early delta-winged aircraft.
Apart from the work of Perkins at Cardington, reported earlier, little thought
appears to have been given to the use of such a configuration. It does have
obvious merit in that all parts of the aircraft could be designed to do work,
i.e. produce lift, whereas the fuselage in most man-powered aircraft is at best
redundant and at worst represents a significant drag and weight penalty.

In addition to the work carried out by Lippisch, Robert Graham, in surveying
the American scene following the announcement of the Kremer prize, found
approximately one dozen feasibility studies being initiated, although interest
in the Kremer Competition as such was minimal. He identified groups at Purdue
University, Massachusetts Institute of Technology (MIT), Georgia Technical
University, Michigan State College, the Bell Telephone Laboratories, Grumman
and the Philco Ford Corporation. Of these, only the work at Georgia and MIT
has resulted in recognisable hardware, but two members of the Purdue University
group, Curtis Cole and John McMasters, presented a paper on man-powered flight
on the occasion of the World Soaring Championships at Marfa in Texas. One
outcome of their survey was a recommendation that an international man-powered
aircraft society be set up to co-ordinate attempts on the Kremer prize and to
disseminate technical information. Support for such an organisation was
lacking, however, which is not surprising in view of the fact that a perfectly
acceptable body, the R.Ae.S. Group, already existed for just these purposes.

The aircraft built at Georgia Technical University was designed by James
M. McAvoy, an aeronautical engineering student. Work began in the Spring of
1962. A low wing monoplane, the main identifying feature was a 2.13 metre
diameter, four bladed ducted propeller located at the rear of the triangular
cross-sectional fuselage. The 26.6 $m^2$ wing had an internal structure of
aluminium and balsa, and was covered with 'Mylar' plastic film, as was the
fuselage, the latter being an aluminium tubular framework.

The contra-rotating propeller was shrouded in the belief that its efficiency
would be increased. This is true, but quite often a study will show that the
slight thrust advantage will be overcome by the increased profile drag caused
by the presence of the shroud, and its weight penalty. In the case of the
McAvoy MPA-1, the shroud also acted as rudder and tailplane, so the latter
argument against its adoption was more or less ruled out. Elevators formed the
top and bottom of the annulus, and the sides acted as rudders. These were to
be used in conjunction with differential flaps on the wing to control right and
left banks. Control surfaces were operated via a single column and hand lever
located in front of the pilot in the open cockpit. One of the principal
advantages to be gained by locating control surfaces around a propeller duct is
that here the air velocity is relatively high, and as a result forces produced
by the surfaces are greater. Thus more manoeuvrability can be obtained, or a
given handling characteristic can be ensured with only small control surfaces
and surface deflections.

Table 13.  Main dimensions of the McAvoy MPA-1

| | |
|---|---|
| Wing span | 16.45 metres |
| Wing area | 26.8 m$^2$ |
| Root chord | 2.28 metres |
| Tip chord | 0.91 metres |
| | (excl. end plates) |
| Fuselage length | 4.88 metres |
| Propeller diameter | 2.13 metres |
| Propeller speed | 240 rev/min |
| Duct diameter | 2.18 metres |
| Take-off speed | 27 km/h |
| Empty weight | 50 - 57 kg |
| AUW | 128.5 kg |

Transmission of power from pedals to the propeller was put into effect using a torque tube, similar to that employed by 'Puffin', with bevel gears at each end.  The torque tube spanned 3.35 metres.

The first attempted flight, during which the machine crashed, was made at nearby Fulton Airport.  Piloted by Robert Richie, who had produced sustained efforts of 0.38 kW on a training rig, the aircraft made a 45 metre dash down the runway, but did not take-off.  During a temporary instability the aircraft keeled over and was badly damaged.  It has since been reported by Sherwin that the project has been taken over by Richie, who is reconstructing the framework.

## THE JAPANESE 'LINNET' PROGRAMME

One of the most significant programmes of man-powered aircraft development to be carried out during the past decade outside Britain, and one which is still being pursued, is that initiated and led by Professor Hidemasa Kimura of Nihon University in Japan.  Since the commencement of the preliminary design study on a machine in April 1963, four prototypes have been built and flown with varying degrees of success.

The construction of the first aircraft, 'Linnet I', *(Fig. 131)* was started in September 1965, and the first flight was made on 27th February, 1966, during which it achieved an altitude of about 2.75 metres, maintaining height for about 9 metres.  The wing was a one piece cantilever structure carrying 104 ribs.  The spar was fabricated using Japanese spruce for the flanges and balsa for the webs.  To maintain the NACA 63$_3$ 1218 section, the wing leading edge was sheathed in 1 mm thick balsa, the remaining surface being covered with styrene paper with a thickness of 0.75 mm.  Wing tips were carved from solid balsa and were used to support skids.  A small amount of dihedral was provided, and the low wing position was chosen to enable full benefit to be made of the ground effect.  A fuselage-cum-tail fin similar to that envisaged by Nonweiler was used to house the pilot, who was in a semi-reclining position.

Spruce was used for giving strength to the front fuselage, with glass fibre reinforcement, and balsa formed the basis of the rear portion.  The whole was covered with styrene paper.  The landing gear comprised two Michelin bicycle wheels, but brakes and shock absorbers were not provided. With a distance from the crankshaft to the propeller of approximately four metres, it was convenient to use a torque tube and bevel gears for power transmission, the tube in this instance being of 'Duralumin'. The propeller was made using a fibre reinforced plastic lay-up formed on balsa ribs and stringers, and had a RAF 6 profile. As a result, it weighed 1.4 kg.

Table 14.  Main dimensions of 'Linnet I'

| | |
|---|---|
| Wing span | 22.3 metres |
| Wing area | 26 m$^2$ |
| Aspect ratio | 18.5 |
| Wing section | NACA 63$_3$1218 |
| Root chord | 1.8 metres |
| Tip chord | 0.6 metres |
| Dihedral | 3 degrees |
| Sweepback | 0 degrees |
| Aileron area | 2.2 m$^2$ |
| Fuselage length | 5.6 metres |
| Tail height | 4.185 metres |
| Tailplane span | 5.4 metres |
| Fin area | 0.97 m$^2$ |
| Cruising speed | 27 km/h |
| Propeller diameter | 2.7 metres |
| Propeller speed | 160 rev/min |
| Aircraft empty weight | 50.6 kg |
| All-up weight | 105 - 110 kg |

The aileron and elevator controls were moved with the aid of a control stick on the right-hand side of the pilot's seat, but rudder control was effected using a handgrip on the left side. Power requirements were calculated to be 0.3 kW at take-off and 0.23 kW for the cruise. The Japanese group obtained a correlation between the power requirement and the aircraft all-up weight of the form:

$$(\text{Power required}) \quad \alpha \quad (\text{All-up weight})^{1.5}$$

This relationship implies that if the weight of 'Linnet I' plus crew had been increased by 20 per cent from 110 kg to 132 kg, this would have required an increase in power of over 40 per cent. This emphasises the need to keep weight to a minimum. During trials to determine the best conditions for continuous high power output, the group found that the optimum pedalling rate was of the order of only 78 rev/min. It was found possible to sustain a power output of 0.26 kW for extended periods at this condition.

It was estimated that 'Linnet I' cost the equivalent of £1000 to construct, and the eighteen student members of the group certainly deserve credit for the speed with which they assembled the machine. Flights of up to 45 metres were made in March 1966, but the group considered that improvements in the pilot position and power transmission could result in a better performance. These modifications were built into the first prototype, which was re-christened 'Linnet II'.

The rebuild of 'Linnet I' was completed late in 1966.  The same wing and basic fuselage was retained but the pilot, who on the first flights of 'Linnet II' was S. Sato, operated in a conventional cycling position.  Also a universal joint was used on the drive shaft, instead of the inefficient bevel gear.  More comprehensive instrumentation was provided in line with the eye level of the pilot, and mounted on a boom projecting forward of the fuselage, as on 'Puffin'. These included an air-speed indicator, angle of bank and angle of attack arms. All-up weight of 'Linnet II' was 100 kg, and the aircraft succeeded during 1967 in making thirty-one flights, the longest of which was 91.5 metres.  An altitude of 1.5 metres was reached, *(see Figs. 132 and 133)*.

Two further Linnets have since appeared.  'Linnet III' first flew in March, 1970, the design study commencing a year earlier.  A new wing section was adopted, and dihedral doubled to 6 degrees.  The wing span was increased to 23.8 metres.  It was not long before this aircraft was modified and redesignated 'Linnet IV'.  This differed from the third prototype in minor detail, including the tailplane section specification.  First flown in March, 1971, the empty weight of 54 kg seems high, bearing in mind the all-up weight of 100 kg achieved with 'Linnet II', an aircraft of almost identical configuration and size.  It is probable that without radical redesign and incorporation of new materials, the Linnet series had reached its limit as far as performance improvements were concerned.

In the mid-1970s the Japanese team under Professor Kimura constructed further aircraft, namely the 'Egret' and the 'Stork'.  The 'Stork' has flown a maximum distance of 595 metres, and in early 1976 yet another machine, using a very intricate balsa-wood structure, was reported to be under construction.

Two other projects of lesser significance were initiated in Japan in the late 1960s.  One, the 'Satoh-Maeda OX-1',(SM-OX), was a monoplane with a wing span of 22 metres and an empty weight of 55 kg.  Construction extended over a period of three and a half years, and the designer was aided in this by a local gliding club.  The first flight took place on 24th August, 1969, and subsequent flights of 28 metres at altitudes of up to 2 metres have been reported, *(see Fig. 134)*.

A slightly smaller but heavier aircraft was that designed and constructed by Eiji Nakamura.  His machine, identified as the MP-X-6, had a wing span of 21 metres and an empty weight of 60 kg.  It was easily recognisable by the wide spacing of the twin tail booms, and was a mid-wing monoplane with a Göttingen 532 wing section.  The pusher propeller was located behind the pod fuselage. After an unsuccessful first flight attempt in 1969, structural failure ending the ground run, no further reference has been reported, *(see Fig. 135)*.

* * *

Recent news from the Soviet Union and other eastern European countries on man-powered aircraft has been notable for its absence.  The early correspondence between MAPAC and the Russians, who were known to have considerable interest in ornithopters, was not actively followed up, and thus the only channel of reliable information was closed.  We can only assume that silence implies little success.

The Polish newspaper, *Skrzydlata Polska*, contained an article in late 1962 recounting the construction of an ornithopter at Lublin/Swidnik Airfield. Designed by an engineer named Beina, who used the theoretical work of a Soviet scientist, Golubiev, this unsuccessful machine had a three-part wing, the outer two sections of which were moveable.  Some faith must have been shown in his design, as official financial aid was forthcoming.

*  *  *

Of the numerous attempts to design and construct a wide variety of man-powered aircraft in the United States, Australia, Africa, the continent of Europe and Japan during the 1960s, only the Linnet series in Japan can be regarded as technically proficient in design and construction, taking as standards such aircraft as 'Puffin I' and 'II', and the Southampton University machine.

It must be concluded that, with the major exception of Canada, discussed in Chapter 12, the extension of the Kremer Competition to international status has not had the desired affect of encouraging the formation of competent groups outside Britain.  This is no reflection on the organisation of the Competition - it is symptomatic of the attitude to and interest in advancement in this field of the potential members of groups, namely aeronautical engineers and engineering students, within these countries.  Perhaps man-powered flight missed, until recently, the contribution of a country which had a particularly viable and profitable aircraft industry, namely France.

*  *  *

*Fig. 125.* S.W. Vine of South Africa standing beside
his partially completed aircraft, which later
crashed.

*Fig. 128.* The Austrian aircraft designed by Josef
Malliga in 1967, about to take-off.

*Fig. 129.* The Malliga aircraft in flight. Note the
negligible dihedral compared with other
machines.

*Fig. 131.* Linnet I, first of a series of four successful aircraft produced at Nihon University.

*Fig. 132.* Linnet II.

*Fig. 133.* Linnet II in flight.

*Fig. 134.* The Satoh Maeda SM-OX, which made several flights.

*Fig. 135.* The Japanese Nakamura MP-X-6.

# 14

# Man-Powered Rotocraft and the Persistence of the 'Bird-Men'

# *14*

Unfortunately, the number of unsuccessful man-powered aircraft greatly exceeds the successful ones, even if one is sufficiently generous as to define 'successful' as being a short flight following rather dubious take-off methods such as running up a ramp.  As we must adopt a much more demanding definition, the count of successful designs does not run far into two figures.

The classification of failures must include two main sections, one being the lack of success following what can be a first rate research and development investigation of the potential of a particular aircraft concept.  In this chapte such basic work on the man-powered helicopter, a layout which has never achieved the popularity of the fixed wing, is discussed.  The ornithopter, it may be argued, would fall within the same category as a man-powered helicopter. However, most of the designers of flapping wing machines have not fully appreciated, or have been unable to reproduce, the motion of a bird's wing.  It is only comparatively recently that work on any scale specifically directed at applying the principle to man-powered flight has been carried out with sufficient scientific objectivity.

A second category, possessing in this instance many interesting examples, is that typified by the amateur aviator as depicted, not unkindly, by the popular press.  Some particularly spectacular and colourful attempts at man-powered flight have been made since the announcement of the Kremer Competition by enthusiasts showing qualities of energy, tenacity and skill in varying proportions.  A small number of these are noted here.

## MAN-POWERED ROTOCRAFT

All truly successful flights made by man-powered aircraft have been carried out using fixed wing machines, conventional monoplanes with the high aspect ratio wings necessary for making full use of the advantages of flying close to the ground.  A number of investigations have been made into the possibilities of constructing a man-powered rotorcraft, and considerable interest in this type of aircraft was generated by the Cierva Memorial Prize Essay Competition of 1958, sponsored by the Helicopter Association of Great Britain.  Contributors were asked to write on the subject:  'Is Man-Powered Rotating Wing Flight a Future Possibility', and significant, if discouraging, papers on the subject were written by Kendall, Naylor, Shenstone and Whitby.

Kendall, who was at the time a member of Saunders-Roe Helicopter Division, concluded that man-powered helicopter flight was just possible provided that the design was kept within very stringent structural and aerodynamic limits. He considered that if a helicopter could be built whose empty weight, excluding blade weight, was about 23 kg per occupant, with a blade weight related to blade area of 2.2 $kg/m^2$, it could be hovered for about thirty seconds.  To do this the blade profile drag had to be no greater than about 0.007.  Comparing this figure with the profile drag of wings on conventional man-powered aircraft, it can be seen that this is a tall order. (Helicopter rotors have similar aerodynamic sections to those of aircraft wings).

The rotor configuration chosen, in particular the number of rotors as opposed to the number of blades, would have a considerable influence on the size and structural sophistication of the aircraft. Helicopter configurations considered by Kendall concentrated on locating the rotors as close as possible to the ground, to make full use of ground effect. The three alternatives suggested are:

(i)   single rotor.

(ii)  side-by-side double rotor.

(iii) double coaxial rotor.

Considerable reduction in rotor diameter could be obtained by using more than two rotors, but drive complexity, Kendall stated, would almost certainly outweigh the gains.

He also investigated the control and stability problems of his man-powered helicopter. To maintain a hovering capability, automatic stabilisation would be necessary. Little work was done on the topic of forward flight of the helicopter, but Kendalls' calculations on the duration of hover under ideal conditions, thirty seconds, make any attempts at flight in a horizontal direction of minor significance!

Naylor's short design study for a single seat man-powered helicopter concentrated on the necessity for forward flight in addition to the achievement of the hover, and his preliminary calculations of drag showed that there was in fact an advantage in designing on a forward speed case, a feature perhaps missed by Kendall. An interesting feature of his design was the adoption of rotor propellers for propulsion, this being one of the few ways for ensuring rotation without reaction, excepting the use of an additional rotor. Naylor conceded that the propulsive efficiency might be of the order of only 70 per cent, but alternative arrangements would probably not increase this figure. The rotor, of 21.3 metres diameter, has a mean chord of 1.53 metres, and thus has a surface area in excess of that of the Southampton University aircraft wing. To minimise the initial power required to take off, a ground run was recommended as a more suitable means for generating lift than an immediate vertical climb.

Some thought was given to the problem of controlling a man-powered helicopter, but Naylor concluded that the three main modes of control, pitch, yaw and roll would be insufficient. He recognised the difficulties in applying control forces at the same time as pedalling, and appreciated that considerable simplification would be necessary. His suggestions for the power drive included full realisation of the torque reaction problem, in which the pilot would rotate! Power from the pedals could be taken to a differential and the two outputs then taken by chain and sprocket wheel to the hub. One output would go to the propeller shaft drive and the other to the rotor hub, rotating in the opposite direction, (see Fig. 136). The torque unbalance could be removed by a cross coupling between the two sides of the differential.

In conclusion Naylor considered that a successful flight could be made under man-power using a helicopter, provided that weight could be kept to a minimum and full use was made of ground effect. The duration of flight,

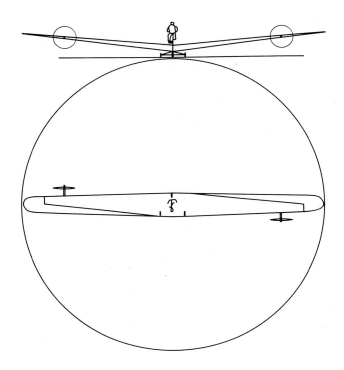

*Fig. 136*  Naylor's proposal for a man-powered helicopter.
Rotor diameter is in excess of 21 metres.

however, would be very short.

The third study on man-powered helicopters was carried out by Shenstone
and Whitby.  They took as the basis of their design the Seehase aircraft,
(Chapter 5) and assumed that the main rotor had the same span, 13 metres, as
that aircraft.  Halving the chord gave a rotor surface area of 6.5 m$^2$ and a
disc area of 134 m$^2$.  An anti-torque rotor was incorporated on a tail arm, as
on conventional powered helicopters, *(see Fig. 137).*

Power calculations based on an all-up weight for a single seat man-powered
helicopter of 111 kg (this weight assumed the use of similar materials to those
used by Seehase) showed that 3 kW was required for hovering flight, and 1.15 kW
for forward flight at a speed of 24 km/h.  It was proposed to use an energy
storage device, in the form of a flywheel weighing about 10 per cent of the all-
up weight, and this would have made a short flight possible.  The flight plan
would be:

   (a)  Store energy for sixty seconds.

   (b)  Engage rotor and leave ground five seconds later.

   (c)  Accelerate and climb to 0.6 metres in three seconds.

   (d)  Decelerate and touch down in two seconds.

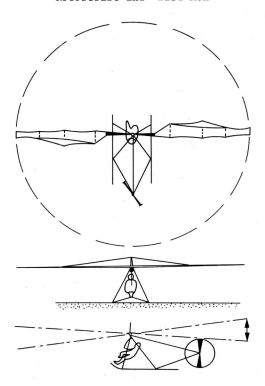

*Fig. 137*  The man-powered rotorcraft proposed by Shenstone
and Whitby, in which they adopted the wing form used by
Seehase in the 1930s.

In the latter stage, appropriate control would ensure flight in a horizontal
direction.  The total flight would be over a distance of about ten metres, and
would last five seconds.  A two-seat helicopter would be able to make a longer
flight, as Shenstone calculated that the weight increase, by a factor of 1.85,
would be more than matched by the corresponding power increase of 2.48 times.

This appears to be Shenstone's sole venture into the field of man-powered
rotorcraft, and he subsequently became a firm advocate of the fixed wing
'conventional' configuration.

A more comprehensive design study for a man-powered helicopter was carried
out by R. Graves of what was then A.V. Roe & Co., in 1961.  He followed up the
work of Kendall, Naylor, Shenstone and Whitby in more detail, in the hope of
showing that their conclusions were too pessimistic.  Graves assumed that a
cruise power of 0.38 kW could be obtained, and also stipulated that the weight
of the rotorcraft should not exceed 113.5 kg of which 45 kg was allowed for
the structure.  Although other workers showed that hovering flight was completely
impracticable with only 0.38 kW available, Naylor found that under forward
flight conditions the reduction in induced power requirements more than
compensated for the increased profile drag power.

As his basis for a rotorcraft which could fly at a forward speed in ground effect, Graves chose a machine which could hover 0.3 metres above the ground and then set out to determine the forward flight characteristics of this basic machine. Four configurations were chosen, having single, double, triple and quadruple rotor arrangements, each rotor being in the same horizontal plane. It was found that higher rates of climb could be obtained by using four rotors, each of 4.7 metres radius, than by utilising a single 20 metre diameter rotor.

Graves then studied the influence of weight on the helicopter configuration. The three rotor structures considered were a light alloy skin, a light alloy channel section spar, and a spruce box section spar. The latter two would have balsa-wood ribs and be covered with light parachute silk. The use of a wood spar as the basis for the structure proved to be the lightest, and the two or three rotor configurations were favoured from the total weight aspect.

For the case of a single rotor, the simple transmission system would be an advantage, but a tail rotor would be necessary to balance the torque of the main rotor. (An alternative method for overcoming the torque problem would be to drive the rotor with propellers near the tips, but this would be inefficient). With contra-rotating rotors, the two-rotor configuration is still attractive in that torque effects are non-existent. This, together with the weight advantage, makes the choice of a double rotor a reasonable solution, as three rotors would again bring in a torque effect, as well as over-complicating the transmission system.

Based on the two-rotor configuration, Graves estimated that his helicopter could be produced within an all-up weight limit of 113.5 kg assuming a pilot weight of 68 kg. The layout of such a machine was extremely simple and was based on a cycle frame minus front and rear wheel forks. These forks were replaced with near-horizontal bars of considerable length supporting fore and aft (tandem) or side-by-side rotors. A double chainwheel was retained for primary drive.

Graves was singularly unambitious in his performance claims for such a machine, should it be constructed. His choice of rotor radius would limit the helicopter to a forward speed of 1.53 metres/sec at an altitude of 1.5 metres on the assumption, of course, that stability and forward motion could be satisfactorily implemented using a straight-forward rotor tilting method. In equating the performance to the requirements laid down for the Kremer Competition, Graves determined that a 7.63 metre radius rotor would be needed to achieve the necessary 3 metres altitude. He felt that such a set of rotors was not totally impractical, and proposed that a five-part study of the major systems could lead to what he believed to be a flying machine having several advantages, not least the ease and convenience with which flight trials could be carried out, compared with a fixed wing aircraft as a means for exploiting man-power. The projects would cover transmission design, rotor blade fabrication, optimum crew complement, ground effect and drag measurements, the last two features being investigated in a wind tunnel.

Most of these very necessary steps in the development of a feasible man-powered rotorcraft design have never been taken, and there is thus considerable potential for work in this field. It remains doubtful whether such a layout could challenge or even approach the performance records set by fixed wing aircraft, however, although another decade will probably pass before we can finally state that this is the case. The solution of the stability and control difficulties likely to be encountered in forward rotorcraft flight are likely to be the key to a breakthrough in this development.

## THE AMATEUR 'BIRD-MEN'

Unlike the professional engineer, who is expected to learn by personal experiences and also by reference to the work of others within his field of activity, the enthusiastic amateur, in planning his design and construction of a man-powered aircraft, would probably rarely consult papers detailing the work of university groups and the like, and would be even less likely to apply correctly any knowledge gained by reading such literature. The reasons for this are many and varied. Access to such information is difficult for a person unfamiliar with the aeronautical engineering industry and the associated professional bodies, although the Royal Aeronautical Society, through the Man-Powered Aircraft Group, has for over a decade provided any enquirer with introductory literature and lists of references. A more insurmountable problem is the lack of technical training of many an earnest flier. A few are able to learn the rudiments of aerodynamics, but this can often complicate the issue, the designer tending to study one point which has attracted his concentration, at the expense of a broader view. Finally, the feeling that one must be right, the certainty of having the correct answer to a problem which has not yet been fully solved, the inflexibility of some amateurs, has considerably inhibited progress within their ranks. One must remember however, that the spirit in which many of these attempts are undertaken is decidedly unscientific, and for that reason, most absorbing!

One of the most publicised individual efforts to design and construct a man-powered aircraft recently was that of Alan Stewart, of Chesterfield. His interest in the subject arose in the late 1950s and in common with many other attempts, considerable impetus was given to his project by the announcement of the Kremer Competition.

Stewart's first man-powered aircraft, 'Belle Bird', was conceived in 1962 and was called after the South American bird of the same name. The aircraft was an ornithopter with a wing span of approximately 6 metres. A considerable amount of research work was carried out prior to adoption of the wing design, and the builder constructed a wind tunnel for basic research on the effect of flapping motion on the airflow over the wing.

'Belle Bird' had an empty weight of 22.7 kg and was constructed from spruce with an alloy framework, nylon being used to cover the upper surface of the wings. The pilot was slung in a tailored nacelle beneath the main plane and his arms were used to control direction by means of a control column hanging from the central fuselage member. Motion was imparted to the wings using the legs through a crank and chain arrangement. The amplitude of motion at the wing tips was designed to be of the order of 0.3 metres, but during tests it was found that the wings were so flexible that they often came into contact with the ground.

Flight tests of the prototype ornithopter began at an airfield near Lincoln in 1964. The aircraft did not take-off under its own power during trials, but towed runs were made carrying ballast, and the pilot experienced noticeable lift generation by the wings, which were beaten at a rate of one cycle per second. Design take-off speed was 24 km/h.

The aim of Mr. Stewart in constructing this first ornithopter was to gain

knowledge and experience which could be applied to his proposed second
prototype, 'Belle Bird II'. 'Belle Bird II' is a much larger machine, the
pilot being in a sitting position driving through pedal cranks and bicycle
chains to the propeller and lower set of moving wings. The upper wings are
fixed. As with the earlier aircraft, the framework is of wood and alloy
construction and weighs considerably in excess of the 22.7 kg of the first
aircraft.

'Belle Bird II' was first tested in 1967, and made a short flight of about
45 metres. Mr. Stewart is still working on this aircraft and hopes to make
longer flights. His aim is to make a sporting man-powered aircraft, capable
of short flights. The work of Otto Lilienthal has had a considerable influence
on his adoption of the ornithpoter concept instead of a fixed wing machine.

* * *

The man-powered 'Cyclopter' designed and built by Warrant Officer Spencer
Bailey, then of R.A.F. Melksham, may be inappropriately placed in this section
and the reader is advised to select his own 'slot', *(see Fig. 138)*.

A two-seat machine, the Spencer Bailey rotorcraft used a 6.1 metre diameter
lifting rotor based on that fitted to the Bensen Gyrocopter. Compared with the
blades of the gyrocopter, the man-powered rotorcraft set had an increased chord
and a thinner profile, except at the root section, where extra stiffening was
incorporated.

Major airframe members were of mild steel tube, contributing substantially
to the empty weight of 47.3 kg. The pilots sat side-by-side, pedalling to
turn a single chain-wheel, the chain leading to a bevel gearbox (modified from
an Avon jet engine unit) which stepped up the drive in the ratio 1.75:1. The
rotor was tilted for directional control by using a phosphor-bronze spherical
bearing as a pivot, which had previously seen service as an Aden gun mounting
component!

The rotorcraft was exhibited at the R.A.F. Art and Handicraft Exhibition at
the Air Ministry in 1962, and observers considered that it lifted from the
floor during a demonstration. Shortly after this, Bailey received a posting to
Aden and no further news of his machine has been forthcoming!

* * *

Many of the less ambitious designs were based on a simple bicycle frame.
William West of West Bridgeford, Nottinghamshire, mounted no less than four
1.53 metre diameter lifting rotors and a pusher propeller on such a frame.
John Davis, a former cycling champion, was trained as a pilot in 1961 for the
pedal-driven helicopter designed by a Hastings artist named Tom Hill. This
aircraft, named the 'Helipede' remained on the ground in spite of the exertions
of Mr. Davis, *(see Fig. 139)*.

* * *

Mr. Barnard of Brixton made a 'Flycycle' using bicycle accessories, in
1966. He was seemingly more optimistic than Mr. West, retaining the four-rotor
configuration but omitting to provide a propeller for forward motion. Total
weight of the machine was 40.9 kg. E. Winter, also of London, limited his

experiments to the testing of a propeller on a bicycle in order to check the
possibility of attaining take-off speed.  Well-wishers christened the
combination the 'Glastonbury Zodiac'.

\* \* \*

The most complex array of bicycle wheels and other components, assembled
in the form of a man-powered helicopter, was produced by Robert Wilson of Preston
in Lancashire in 1965, *(see Fig. 140)*.  Prior to this, W.L. Manuel designed and
manufactured a very neat fixed wing monoplane mounted onto a bicycle - a much
more professional effort than some of the early 'Aviette' designs in the 1912
and 1913 Peugot competitions, *(see Fig. 141)*.

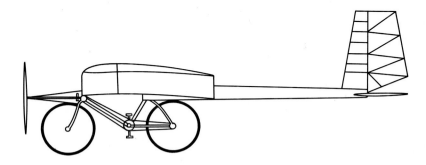

*Fig. 141*  A view of the Manuel 'flying bicycle'

A more crude winged bicycle was made by Barry Crocker in Leeds in March 1966.
With fixed wings the empty weight was 81.8 kg, and the wing area to support
this only 5.57 m$^2$.  A tractor propeller was included.

\* \* \*

Direct exponents of bird flight are few now, but the attraction of
publicity encourages some to amuse us with their antics.  Don Partridge, well-
known in the early 1970s as a popular entertainer, once jumped hopefully off
Hammersmith Bridge into the Thames, carrying a pair of wings strapped to his
back.  In 1970 Walter Cornelius, a Latvian by birth, launched himself in
similar manner from the 9 metre high roof of a supermarket backing onto the
River Nene at Peterborough.  This 'dive' set a fashion, and the owner of the
supermarket has offered a prize to any person who can fly across the Nene,
using wings of a certain maximum size.  In 1971 numerous attempts were made by
apparently sober citizens of Peterborough, including a mass attempt by
schoolboys, which was spectacular in its failure!

\* \* \*

A number of largely unsubstantiated claims regarding flights were made in
the 1960s.  A short flight was claimed by a Mr. Watts of Coventry using an
ornithopter.  Models had been constructed prior to this in an attempt to
produce a satisfactory design, and several had flown.  Mr. Walton of Taunton,
supporting a machine weighing 45 kg, crashed after a take-off run lasting
20 seconds.  Douglas MacDonald constructed an aircraft designated MAC-2 during

the period in which he was employed by the old Beagle-Auster Company at Rearsby, Leicestershire, *(see Fig. 142)*.

* * *

Other projects which received attention at the time included a proposal for a man-powered flying saucer, the idea of Roy Bryant of Earls Court, London. Helicopters were designed by Messrs. Ways and Dyer, and Graham Rouse of Tichfield, Hampshire, and Leslie Hesse of Hadleigh, Essex made a monoplane of considerable dimensions, based on a tricycle framework.  A university 'Group' which could not provide competition for SUMPAC was represented at Lancaster by John Lavery, who built a 7.3 metre span monoplane.

* * *

Contributions to the English language were the most noteworthy features of the proposals dated 1964, of Ian Child.  He suggested a man-powered 'Minicopter' which had a fifteen minute endurance and a cruising speed of 64 km/h.  Flapping 'dynofoils' were to be provided on each side of the fuselage, in addition to conventional wings and tailplane.

* * *

R. Westmacott of Swindon, designed a man-powered helicopter while under-taking an aeronautical engineering apprenticeship.  His engineering skill enabled him to construct a machine weighing only 8.2 kg, but rotor diameter restrictions proved too great, Westmacott calculating that an 18.3 metre diameter unit would be needed.  His frame was only suited to a rotor one half this size.

* * *

Two other 'attempts' were made in Britain around this time.  Leslie Smith was seen with his biplane at London's Heathrow Airport, *(Fig. 143)*, and Clifford Davis amused the local children with his 'helicopter' in 1963, *(see Fig. 144)*.

* * *

Herbert Watkinson of Bexhill-on-Sea, claims to have made the only successful man-powered autogiro flight to date.  Having developed an interest over a number of years in ultra-light and man-powered aircraft, he commenced experiments on aspects of autogiro construction in 1961, and in June 1962 claimed a flight of between 35 and 55 metres.  This was witnessed by two observers.

The Watkinson machine had a two-bladed rotor of slightly over 9 metres diameter, with detachable blades to ease storing and transport.  Forward propulsion was obtained using a 1.14 metre diameter airscrew.  With an empty weight of 34 kg, the machine was very heavy and rotor loading was high. Flapping hinges were provided to permit the rotor blades to take up their preferred positions and give uniform lift, and in details such as this the design was quite advanced.

In 1970 the Shuttleworth Trust arranged to accept the Watkinson autogiro and it was hoped that they would carry out trials.  However, no more has been heard of this machine.

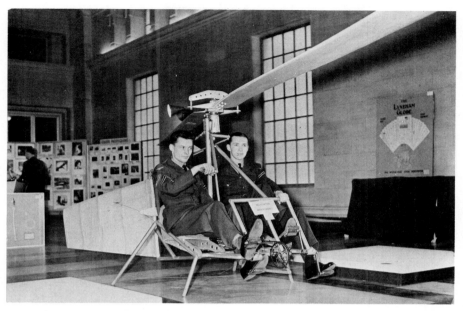

*Fig. 138.* The Spencer-Bailey rotorcraft on exhibition. Witnesses saw it rise a few centimetres when tested.

*Fig. 139.* The John Davis 'Helipede'.

*Fig. 140.* Robert Wilson of Preston, Lancashire, made
this contribution to industrial archaeology
in 1965.

*Fig. 142.* D. MacDonald, of Leicestershire, was one
of the many who failed to 'rise to the
occasion' in the 1960's.

*Fig. 143.* Leslie Smith, of Middlesex, constructed this biplane in 1967 with a view to entering the Kremer Competition.

*Fig. 144.* Clifford Davis - an amateur inventor who constructed this machine in 1963. Unfortunately the transmission system does not appear to be up to scratch!

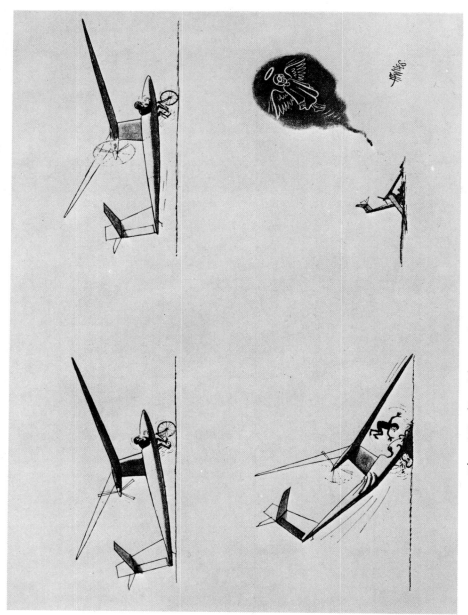

Fig. 145. 'Punch' magazine made one of its usual apt comments on this activity.

# Research
# and
# Development
# Work

## — The Basis for a New Generation of Aircraft

# 15

Much of the progress made in the field of man-powered flight in the 1960s has been as a result of comparatively costly and complex design and manufacturing exercises with, in some cases, timescales of several years. In general the most important work has been carried out by groups, as opposed to individuals. It has been the engineering departments of universities and the aircraft industry who have provided the forum for these endeavours.

Possibly one of the most significant moves in the pre-war era of man-powered flight was the setting up of the Muskelflug-Institut by Ursinus and others in Germany. If circumstances had permitted the proposed research and development programmes of this organisation to be carried out, much of the knowledge gained since 1959 could have been available to us twenty years earlier.

## RESEARCH AND DEVELOPMENT PROGRAMMES

It was at the College of Aeronautics, Cranfield, that the reawakening of interest in practical terms, in man-powered aircraft occurred, and the formation of MAPAC, described in Chapter 8, was to provide a basis for valuable research work, in addition to encouraging programmes directed at producing a complete aircraft as an end product. It is perhaps sad that the work pioneered at Cranfield did not result in the establishment of an English equivalent to the Muskelflug-Institut[1]. As subsequent events have shown, co-ordination of projects of the type that individual researchers, or small teams, could carry out has been lacking if only because the Royal Aeronautical Society has not been the sole controller of the purse-strings. With individual university departments and the Science Research Council additionally providing funds, duplication of work – also to be expected because of the competitive nature of such a venture – and the selection of less important topics for research, was inevitable. However, the universities and some individuals have taken part in, and are still carrying out, programmes which contribute in some fair measure to our aim of achieving the 'ultimate' man-powered aircraft. Appropriately, it was the College of Aeronautics that 'set the ball rolling'.

---

[1] *At the 1970 A.G.M. of the R.Ae.S. Man-Powered Aircraft Group, it was proposed that an attempt be made to provide common hangarage and flying space for all man-powered aircraft. Cranfield was suggested as a possible site for this flying ground, but the geographical location of groups makes such an idea difficult to implement.*

## LILLEY AND SPILLMAN

Nonweiler had carried out a full design study of an aircraft with the eventual aim of manufacturing a prototype, and this proposal is described in Chapter 8.  Later, in 1960, Lilley and Spillman considered the problem from a more objective standpoint, producing an unpublished survey of the design and performance of man-powered aircraft.  They restricted much of their effort to fixed wing types, rightly assuming these to be representative of the simplest form of aircraft and most likely to be successful as an 'early generation' man-powered machine.

Many topics affecting the performance of a man-powered aircraft were discussed by Lilley and Spillman and were grouped in the following major sections:

> Power Generation
> Type of Aircraft
> Main Design Features of a Fixed Wing Aircraft
> Performance in Sustained Flight
> Optimisation of Wing
> Taxi-ing and the Ground Run
> Power Storage.

Wilkie's work on power outputs of a man *(see Fig. 146)*, was used as the basis for calculations, and having fixed the general configuration of the aircraft, this being a high wing monoplane with one or two crew[1] (to avoid mechanical complexity), the primary features of design were investigated in some detail.  Crew weight was the first topic dealt with, and the significance of the weight of the pilot(s) as a proportion of the all-up weight was given emphasis by the authors - as the weight of the crew can be some 67 per cent of the all-up weight, a 10 per cent saving in structural weight only represents a saving of 3 per cent in the all-up weight.  Using a single crew weight within the range 63 kg to 76 kg, and including data on a number of aircraft designs, including the Haessler-Villinger machine, Lilley and Spillman concluded that the overall weight of a single seat aircraft would be approximately 122.5 kg, and that of a two-seater, 250 kg.  If "considerable structural ingenuity" were used, they estimated that these figures might be reduced to 90 kg and 180 kg respectively.  It is interesting to see how developments during the subsequent decade have tended to support their initial assumptions, but it must be borne in mind that wing spans of single and two seat machines are now up to fifty per cent higher than the upper limit suggested in this study.

Lilley and Spillman obtained correlations for wing weight and extra-to-wing weight, based on the Haessler-Villinger aircraft, this being considered representative of a single seat machine, and the Nonweiler design study.  At the time of compilation of the report, information was given by Perkins concerning the weights of his inflatable wing aircraft.  These values fell outside the correlations based on the Haessler-Villinger and Nonweiler machines,

---

[1] *The introductory section concerning aircraft layout included ranges within which the span, aspect ratio and wing area might lie.  Spans of "between 40 ft (12.2 metres) and 80 ft "(24.4 metres) were suggested for the single seat aircraft, and limitations to increases above 24.4 metres were thought to be wing flexibility, high profile drag and difficult control.*

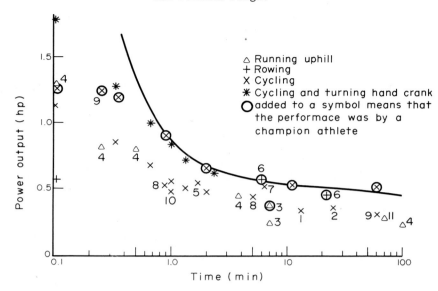

*Fig. 146* The power output of a man, as presented by Wilkie as a function of duration of effort. Note the effectiveness of combined pedalling and hand cranking.

and illustrates the very conservative nature of the weight assessments made. It must be remembered that the Haessler-Villinger 'Mufli' was constructed almost thirty years before this study was completed. The authors did emphasise the fact that their assessment was based on the use of woods such as birch ply and spruce, and that by using foam plastic or honeycomb structures, a very different order of wing weight may be arrived at.

In rounding off the section on weight assessment of man-powered aircraft, Lilley and Spillman succeeded in obtaining a very simple, and for that reason rather approximate, formula relating wing weight ($W_w$), weight of each crew ($W_m$) and the all-up weight of the machine (W), to the number of crew (N). This was of the form:

$$W - W_w - N\,W_m = 30\,(N + 1)$$

where all weights were measured in pounds.

They were among the first to suggest that a child-driven machine might have advantages over that powered by a fully-grown man weighing of the order of 64 kg. Assuming that their correlations for weight and power requirement stood for such a machine, Lilley and Spillman estimated that for a child weight of 34 kg a required power of 0.45 kW in an adult-driven machine would be reduced to 0.24 kW. Using the formula describing the power output of a human in terms of his weight, power $\sim W_m^{0.73}$, if the adult available power were 0.34 kW the available power from the child would be 0.23 kW. Thus whereas in the former case, a man-powered machine, the power required would be greater than that available by 16 per cent, the child-powered aircraft would be under-powered by only 3 per cent. This, it was stated, pre-supposed that the actual wing weight for a given span and area would be reduced to correspond to the saving in all-up weight, this reduction amounting to about two-thirds. A further reduction in power required would be obtained if the extra-to-wing weight were also reduced in proportion. It was suggested that, assuming the wing was designed on the basis of minimum material thickness, the wing weight would be proportional solely to its area, or the specific wing weight (weight per unit area) would be independent of span and aspect ratio. Although less favourable than the design based on the main component weight correlations obtained by Lilley and Spillman, a machine with the features given in the table below could be produced, with the condition that the child would be equally as skilled as an adult pilot.

Table 15. Lilley and Spillman child-powered aircraft.

| | |
|---|---|
| Wing loading | 4.88 kg/m$^2$ |
| Wing weight | 22.7 kg |
| Wing area | 13.94 m$^2$ |
| Lift coefficient | 1.0 |
| Drag coefficient | 0.03 |
| Aspect ratio | 20 |
| Propulsive efficiency (incl. propeller) | 76% |
| Cruise power required | 0.235 kW |
| Power available | 0.228 kW |

In establishing performance criteria for man-powered aircraft, Lilley and Spillman first reviewed the operating height. Assuming still air conditions and no benefit from thermals, power output would theoretically limit the altitude of a single seater to between 9 and 12 metres, (pedalling being the sole power generation technique), but practical considerations would suggest a figure nearer 6 metres. For a multi-man crew aircraft, the restriction on altitude would not be so severe. At this height, indeed at heights up to those given by a height to span ratio of 0.75, some reduction in the induced drag would be anticipated, and the reduction would be particularly strong at 5 to 6 metres altitude, with an aircraft of approximately 25 metres span. A clear warning was given concerning the altitude at which manoeuvres could be safely performed. With skilled crews the authors felt that turns within the above altitude range could be possible, but reference was made to damage incurred by Haessler's wing tip when flying at an altitude of 1.8 metres with a span slightly over half the maximum considered in their study.

No full design resulted from the survey of Lilley and Spillman, but they were able to show that, based on the use of conventional materials and structural layout, man-powered flight with a fixed-wing monoplane was possible, if marginal. Careful design was necessary, and excellent aerodynamic performance coupled to a stiff, light structure were pre-requisites of success, as was the fitness and high athletic ability of the crew, regardless of number. Power storage would, they argued, considerably increase the chances of success, particularly in reducing the fatigue of a take-off and climb manoeuvre. In their concluding remarks, Lilley and Spillman put a damper on the aspirations of the amateur engineer who felt that he could win the Kremer prize, using a pair of wings made in his garden toolshed:

> "It seems very clear that the era of the flying bicycle is not just round the corner. The kind of machine we have contemplated is a very superior type of aircraft and is not the kind of machine which could be assembled in a backyard and flown by anyone from the local sports field by merely jumping onto the bicycle seat and then pedalling flat out. That day may come eventually but it is obvious that much research and development will have to have been accomplished before it can be regarded as even a very vague possibility. Man-powered flight today presents a challenge, and will remain so for some time to come."

This remains true today, and it is interesting to pause and consider that if a full-scale aerodynamically sophisticated man-powered aircraft weighing only 0.5 kg could be produced, the all-up weight would be reduced by little more than 25 per cent compared with what is currently structurally possible in terms of single seat aircraft weight.

## WICKENS — PROPELLER RESEARCH

The work of Lilley and Spillman, like that of other members of MAPAC, indicated to a much wider field of specialists interested in the subject of man-powered flight the type of research and development work needed in overcoming difficulties which could delay and even completely negate the aspirations of those working on the design and construction of actual aircraft.

R.H. Wickens, then of the National Aeronautical Establishment in Canada, wrote an important paper on the topic of propeller selection, his work being published in the Canadian Aeronautical Journal in 1961[1]. From considerations of simple propeller theory, the maximum propeller efficiency is obtained with a low disc loading, or thrust to swept area ratio. Naturally, the best way to ensure a low disc loading is to increase the propeller diameter. This in practice leads to a drop in efficiency, because the total blade profile drag increases, and viscous losses continuously increase as one is using more blade surface. Power is therefore required to overcome these factors before useful thrust is created.

In designing a propeller for a man-powered aircraft, Wickens chose as typical a two-seater with an all-up weight of approximately 227 kg. The most suitable propeller was found to be one with three blades, these having a Clark Y profile (blade profiles being described in the same way as wing profiles which typically have NACA sections). With an optimum calculated efficiency of 88 per cent and an overall efficiency of 80 per cent, it would give a cruising speed of 8.37 m/s. Tip speed would be approximately 15 m/s.

Wickens carried his optimisations a stage further in order to obtain the maximum propeller diameter for a given number of blades. The effect on efficiency of a change in the number of blades was not as great as one might

---

[1] *R.H. Wickens. 'Aspects of Efficient Propeller Selection with Particular Reference to Man-Powered Aircraft'. Canadian Aeronautical Journal. Vol. 6, pp 319 - 330, November 1961.*

expect, but considerable savings in diameter could be obtained. A four-bladed
unit for the 227 kg aircraft would have a diameter of 1.675 metres. However
reductions in diameter produced lower blade section Reynolds numbers and a
lowering of the efficiency.

The propeller efficiency is of course important, and several techniques have
been investigated for minimising blade losses, in the same way that various
forms of boundary layer control may be used on wings to reduce drag and
increase the lift to drag ratio. The first and most obvious way to ensure high
propeller efficiency is to choose a low drag section, as one does with wings.
Laminar flow sections for propellers have been used for some considerable time,
and their use has overcome the high drags normally associated with a cambered
profile. Cambered sections have long been desirable because of their increased
static thrust at take-off.

In reviewing the various methods used in an attempt to improve propeller
efficiency, Wickens noted the work on utilising the centrifugal force of the
blade to provide sucking and blowing boundary layer control. Convection
currents caused within a hollow blade with an intake at the root, exhausting at
the tip, would effect removal of the boundary layer by suction at the root,
and blowing at the tip. It was felt by several workers that only very low
forces would be generated, insufficient to significantly improve performance,
and to date this system has not been tried on a man-powered aircraft. A
second largely unproven method for improving performance is the use of leading
edge slots, most advantageous in the take-off condition in that root stalls
can be avoided.

One layout suggested by Wickens and adopted by McAvoy for the Georgia
Technical College aircraft is the ducted propeller. The increased efficiency
possible with a well-designed ducted propeller system appealed to Wickens in
that the propeller could take over all, or at least a significant part of,
the lift generation, as well as providing thrust. A shroud around the blades
reduces tip losses by acting as an end plate, but in addition the pressure
distribution of the duct itself contributes towards thrust, theoretically
offering a 26 per cent gain in static thrust over an equivalent free propeller.
(This gain is, however, minimised at any reasonable forward speed).

## WILSON — TRANSMISSION SYSTEMS

Experiments at Oxford University by S.S. Wilson clarified the position
concerning the performance of the other major contributor to drive efficiency,
the transmission system. Wilson constructed a test rig which could cope with
the peculiar characteristics of man-powered aircraft transmissions. Their low
speed and heavy peak torque lie largely outside the main operating regime of
most power transmission systems, and the efficiency differences between the
quite wide selection of possible drive systems implies that any useful rig must
be able to measure performance extremely accurately. Power recirculation was
used by Wilson in obtaining accurate recordings, the loss torque being applied
and measured using an air-lubricated bearing arrangement. For positive drives,
a known torque could be locked into the system, and for belt drives, where
inevitably there is a difference in speed between the input and output shafts
due to slip and creep, a known percentage speed reduction was provided by means
of pulleys of slightly differing diameter such that the torque when running was
accordingly self-adjusting.

A wide range of transmission systems, based largely on pedalling as the chosen input method, was tested by Wilson. He investigated, in addition to their direct efficiencies, the stresses imposed on the supporting structure, the effect of structural deflections on the drive performance and the changes brought about by cyclic variations in the torque, caused both by slight power changes during the pedalling stroke and by the thrust changes as the propeller blades pass through wakes of fins, pylons and any other appendages.

By early 1961 Wilson had tested such drives as chain and sprocket, toothed flat and 'Vee' belts, spur and bevel gears, universal joints and double crank and push rod units. He also considered the use of a long thin shaft used in a circular arc, being most appropriate to aircraft with tail-mounted or pylon propellers. The system would be designed such that the shaft ran at a speed less than its primary whirling speed. This would ensure that the natural frequency of torsional oscillation was well below the frequency of the primary component of oscillating torque, which corresponds to twice the pedalling speed. Wilson claimed that this condition would help to reduce cyclic variations in propeller speed, hence slightly increasing efficiency.

Table 16.  Transmission system properties

| System | Efficiency % | Weight kg/m | Load on Structure | Sensitivity to Deflections | Ability to Reduce Cyclic Variations |
|---|---|---|---|---|---|
| Chain | 98.5 | 0.611 | Minimum tension | Shortening | Yes |
| Belts: | | | | | |
|   Flat nylon | 97 | 0.224 | Large tension | Small | Yes |
|   Toothed | 99.5 | 0.395 | Minimum tension | Shortening | Yes |
|   'Vee' | 97 | 0.14 | Medium tension | Small | - |
|   Plain steel | 99 | 0.58 | Large tension | Shortening | Yes |
|   Perforated<br>     steel | - | 0.566 | Small tension | Shortening | Yes |
| Gears: | | | | | |
|   Spur | 97 | - | Direct thrust | Twisting | - |
|   Bevel | 94 | - [1] | Direct and side thrust | Any deflection | No |
|   Worm | 89 | - [1] | End thrust | Twisting | No |
|   Roller Cam | 99 | - [1] | End and side thrust | Small | Yes |
| Hooked Joints | ? | - [1] | Small | None | No |
| Double Crank | 96.5/96.5 | - [2] | Large thrust and tension | Small | No |
| Flexible Shaft | 99 | 0.189-0.45 | Small bending moment | None | Yes |

[1] Duralumin shaft   0.153 kg/metre
[2] Tubes             0.229 kg/metre

The results obtained by Wilson are summarised in Table 16.  Drawbacks of the chain include the fact that it is difficult to operate with even a small twist over its length, and that they tend to be heavy, although lightweight ones are available at a price.  Gears suffer from the disadvantage of being unable to damp out cyclic variations in thrust.  The 'Puffin' used a combination of bevel gears and a magnesium alloy tube split by universal joints towards each end.  This is currently one of the most popular drive systems for man-powered aircraft.

It was hoped that Wilson would have been able to provide a service in testing proposed transmission systems on his rig, but in 1965 he exchanged correspondence with Shenstone, who was writing on behalf of the R.Ae.S. Man-Powered Aircraft Group, in which he stated his inability to progress further on efficiency measurements.  The suggestion that another researcher might take up the work does not appear to have been pursued.

## REAY — MULTIPLANES AND THE MAGNUS EFFECT

Most man-powered aircraft have been fixed wing monoplanes, although invariably designs have been dictated to a greater or lesser extent by the layout of contemporary mechanically powered machines, as explained by Lilley and Spillman.  The influence of glider design, characterised by the use of a single high aspect ratio wing, has in general outweighed other layouts, and even in the 1930s, when biplanes were in vogue, the Bossi-Bonomi, and particularly the Haessler-Villinger aircraft were modelled on gliders.  (Bossi did, however, briefly consider a biplane).

Thus, apart from a few designs based on helicopters, which have not radically changed in concept over the years, little thought has been given in serious aeronautical circles to adopting what may appear to be very unconventional layouts or lifting surfaces for man-powered aircraft.

Towards the end of 1965 the author commenced working on a research programme at Bristol University associated with the examination, both experimentally and theoretically, of more unconventional layouts and lift-generating devices.  The arguments supporting a case for this type of investigation are the following.

It is convenient to break down the drag of an aircraft, or aircraft wing into two major components.  These are the profile drag and the induced drag. The profile drag is largely a function of the shape of the wing, and induced drag depends on the lift coefficient of the wing and aspect ratio.

The equation for the drag may be manipulated such that the criterion for total drag, i.e. the sum of induced drag and profile drag, may depend on the relationship between these two drag components.  In reality, the minimum power condition occurs when the induced drag coefficient (the coefficient being simply a convenient non-dimensional form of expressing a force) is three times as large as the profile drag coefficient.

The induced drag coefficient $C_{D_i}$ may be written as:

$$C_{D_i} = \frac{kC_L^2}{\pi A}$$

where  k  is the induced drag factor,  $C_L$  is the lift coefficient and  A  the aspect ratio.  This expression, under the condition for minimum power, must equal three times the profile drag coefficient, $C_{D_O}$.

Equating these two quantities and rearranging to obtain the lift coefficient in terms of profile drag, the following equation is obtained:

$$C_L = \sqrt{\frac{\pi A}{k} \cdot 3 C_{D_O}}$$

If  k,  the induced drag factor, is assumed equal to unity, and the geometric aspect ratio,  A,  and  $C_{D_O}$  are taken as 22 and 0.01 respectively, the lift coefficient for minimum power would apparently be 1.4, which in itself is high for a conventional wing.  However, if account is taken of ground effect, where the 'effective' as opposed to 'geometric or physical' aspect ratio of the wing may be up to two times as great as the value of 22 used above, it can be seen that by substituting a higher value of  A  in the above equation, a value of $C_L$ as high as 2.0 may be obtained.

This suggested two lines of research, both of which the author was able to follow through in some detail:

   (i)  Research into wing sections or other devices capable of generating high lift without high profile drag, and also without demanding too much in the way of manufacturing accuracy.  (This to some extent may rule out a sophisticated laminar flow wing, although this topic is expanded later).

   (ii)  An examination of ground effect, with a view to reducing drag, related to less conventional wing arrangements.

This latter category can cover a multitude of wing arrangements, and it was necessary to restrict the choice to two major systems, excluding the monoplane, which was taken as a datum.  The arrangments chosen were the biplane and the tandem wing.  (A tandem wing aircraft has two wings mounted one in front of the other).

The research carried out under the terms of reference in paragraph (i) above was mainly experimental in nature.

Design of high lift, low profile drag wing sections is a very sophisticated exercise, and although the author has more recently been associated with attempts to do this, the work carried out at Bristol was devoted to examination of a less well known lifting device, the rotating cylinder.

The rotating cylinder as a lift-generating device has been recognised for many years, but applications have not been very forthcoming.  Possibly the best known use of these devices was as a replacement for sails in the 'Flettner Rotor' ship.  Such was the claimed efficiency of these devices that the sail area of a sailing vessel could be drastically reduced by substituting two rotating cylinders, with their axes vertical.  Such a ship, using mechanically rotated cylinders for forward propulsion, was constructed and sailed reasonably successfully.

Suggestions for applying the Magnus effect - the principle whereby a rotating cylinder generates lift - to aircraft appeared as long ago as 1925, when a considerable amount of research work was carried out in Europe, notably Germany.

Although testing of cylinders operating directly as wings was never successfully carried out, data from simple tests on a single cylinder was applied in
calculations of aircraft performances and size, and some interesting
conclusions reached.

The reports[1] did not cover in detail such aspects as manoeuvrability and
the gyro effect of rotation, but considering power requirements and the lifting
capabilities, it was shown that the rotating cylinder wing could well be used
in cases where a comparatively low power to weight ratio could be substituted
for short take-off and low speed capabilities.  These documents were written
in 1925 and no doubt the advent of the helicopter filled the gap which a
rotating cylinder machine could have occupied.

It was partly the results of work such as that above which influenced the
author in choosing the Magnus effect as a likely candidate as the lifting
system of a man-powered aircraft.

The values of the lift and drag coefficients for two-dimensional (as
opposed to three-dimensional representation, which more accurately models an
actual wing) rotating cylinders are well documented.  No known work had been
done on three-dimensional flow over cylinders however, and the peculiar case
of a cylinder in ground effect, simulating its use on an aircraft flying very
close to the ground, had not been considered.  Some of the results of the
experiments carried out at Bristol University are shown in *Fig. 147*.  By
comparing the values of lift coefficient obtained, with those for conventional
aerofoils recorded in other chapters, the considerable improvement can be seen.
The lift of a cylinder increases with increasing V/U, the ratio of the speed

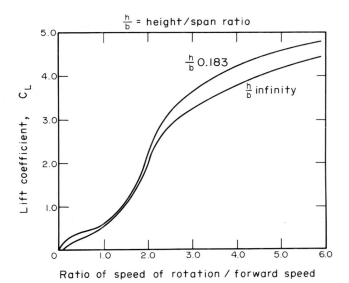

*Fig. 147*  The Magnus effect - variations in lift coefficient
of a rotating cylinder with a ratio of rotational/forward speed.

---

[1] A. Pröll.  *'Consideration of Rotor Problems'*,  *Zeit. Flugtechnik und Motor-
luftschiffahrt, Vol. 16, No. 3, 1925.*

of rotation of the cylinder to the forward speed of the aircraft. The major
drawback however, appears to be a high drag, increasing with increasing V/U.
Ground effect does have a favourable affect on the induced drag but values of
the drag coefficient do remain rather large.

If some of the experimental results are used in the calculation of the
power, using the equation:

$$\text{Power} = \text{drag} \times \text{velocity}$$

the following results are obtained:

For an aircraft with an all-up weight of 122.5 kg and two cylinders
12.2 metres long and 0.914 metres in diameter (these being realistic sizes,
based on the lift required to support such a machine) the power necessary to
fly at a forward speed of 6.4 m/s with the cylinders rotating at 400 rev/min
would approach 2 kW. When remembering that an average fit man can produce a
power of approximately 0.26 kW, the advantages of the apparent high lift
coefficients obtainable using the Magnus effect are more than balanced by the
increase in power required to rotate the cylinders and overcome additional drag.
(The value for the all-up weight, 122.5 kg, assumed at the time is by todays'
standards, rather high. If it were possible to use the form of construction
pioneered by Perkins - an inflatable cylinder - together with a light-weight
transmission system, at the same time maintaining sufficient cylinder rigidity,
this figure could be reduced, saving power. It is doubtful however, whether
the power requirement could be brought within the capabilities of a man at this
stage.)

The investigation of the drag of less conventional wing arrangements,
although purely theoretical in nature, produced more encouraging results.
Reductions in the induced drag of a monoplane are very noticeable in ground
effect, as detailed in Chapter 7, *Fig. 74*.

It can be shown that by using calculated values of the induced drag of both
biplanes and tandem wing arrangements in ground effect, considerable advantages
may be gained over the power required to fly an 'equivalent' monoplane. (In
these calculations it was assumed that an 'equivalent' monoplane was one which
had the same total wing area as the multiplane with which it was being compared)
The advantages proved more marked when flying at an altitude equivalent to about
one fifth of the wing span of the machine, rather than at altitudes closer to
the ground, below the minimum height laid down in the rules of the Kremer
Competition.

The higher span ratios, i.e. multiplanes where one wing was of a noticeably
greater span than the other, appeared to offer the best reductions in drag, but
the ultimate test of this will only be found when allowance is made for the
slightly greater weight involved. It was also indicated that a more favourable
performance would be obtained in the case of a biplane by using wings with
large gaps and staggers. (The gap is the vertical distance between wings, and
stagger describes the horizontal displacement of one with respect to the other).
Structural problems may occur with these configurations, however.

Of course it is not truly representative of a full scale aircraft to
compare powers using only profile and induced drags of wings. A number of
additional contributory factors which together may prove prohibitive by
completely negating any biplane advantages must be considered. Among these

factors are struts between the wings and the bracing wires needed.

A man-powered aircraft utilising tandem or biplane construction layout has not been built in Britain making full use of current technology, and in view of the reduced wing spans possible, it was felt that it might be well worth considering the adoption of such a layout, if only from the point of view of manoeuvrability.

\* \* \*

Three other projects are worthy of note in this Chapter; the design study carried out at the University of Salford, J.S. Elliott's work on the 'fluttered' wing at RAE, Farnborough and another ornithopter project, this being at Bristol Polytechnic.

## SALFORD UNIVERSITY DESIGN STUDY

The Salford project is of interest not least because it sports three crew members. A wind-tunnel model, illustrated in *Fig. 148*, shows the twin boom design, although first conceptions of the aircraft suggested a delta wing layout. Main design parameters finally adopted are tabulated below.

Table 17. Main design parameters of the Salford man-powered aircraft.

| | |
|---|---|
| Wing span | 30.5 metres |
| Wing area | 58 $m^2$ |
| All-up weight | 313 kg |
| Empty weight | 122.5 kg |
| Aspect ratio | 16 |
| Wing loading | 4.9 $kg/m^2$ |
| Cruise lift coefficient | 0.64 |
| Maximum lift coefficient | 1.2 |
| Cruising speed | 40 km/h |
| Stall speed | 30.5 km/h |
| Fuselage length | 10.35 metres |

Work on the project began in 1964, but following the completion of the feasibility study and the construction of a small number of test specimens, the emphasis was switched to a single research project on the behaviour of a monoplane in ground effect, initiated by Dr. H. Portnoy. The University found that the construction of a full-scale machine would have been prohibitive from the point of view of cost and time, and most of the original activity was carried out by undergraduates.

Salford had not determined the 'optimum' number of crew mathematically, but believed three to be a reasonable number as it was still possible to maintain a wing loading as low as 4.9 $kg/m^2$ with a conventional structural configuration. Initially, using the argument that full mental ability would

be required during critical manoeuvres at the same time as peak physical
output, and that such a combination was unattainable with a single crew, a
two-seater was the obvious choice. However, during an early stage in the
feasibility study it became apparent that an extra crew member could be
carried with very little modification to the design. Hand cranking by the
two non-controlling crew members was not considered as it was felt that the
oxygen available for conversion to energy could be more than absorbed by the
leg action alone.

## ELLIOTT — THE 'FLUTTERED' WING

    H.C.H. Townend, writing in the Journal of the Royal Aeronautical Society in
1937, first mooted the idea of man-powered flight using a wing maintained in
a state of propelling flutter, in which the wing provides both lift and forward
propulsion forces[1]. David Rendel, who had been the Secretary of MAPAC and was
at the same time working at the Royal Aircraft Establishment, Farnborough,
commented in 1959 on the high propulsive efficiencies possible on condition
that forward speed is kept low and the correct flapping motion was simulated.
Another advantage given was that these wing motions could result in higher
lift to drag ratio values than possible with fixed wing aircraft, owing to
delays in the build-up of the induced portion of the drag.

    A long-term programme was commenced, being largely the effort of J.S. Elliot
who, until his recent retirement, was an aerodynamicist at RAE Farnborough,
investigating the feasibility of flapping flight, more particularly the
'fluttered' wing. Elliott produced his first ornithopter designs in the early
1960s, and the earliest machine showed the pilot supported on a tricycle
propelling two wings using his arms. The wings were provided with rubber
bungees to aid flapping. Wing motion was predominantly of an up-down form,
but some forward movement was superimposed on the downstroke. Wing span was
only a few metres and the wing planform resembled a conventional system,
having straight leading and trailing edges and zero taper. A standard aerofoil-
type section was proposed.

    Four months later Elliott produced the design of a second machine which
exhibited a similar propulsion system to his earlier model. A lever arrangement
was used to aid wing actuation, with the arms remaining down by the pilot's side
instead of, rather uncomfortably, level with his head, in the Mk. I aircraft.
The pair of wings were joined using a flexible coupling above the pilot in an
attempt to reduce end losses and offer a fixed wing contribution to the lift in
cases where wing actuation was not occurring. An innovation in this second
design was a triangular tailplane controlled by arm levers. Rear wheel drive
from pedals was a feature, and wing planform and construction differed in some
detail from that of the first aircraft. A bat-like trailing edge, and projectio
of the wing ribs above the covering, which was applied only to the wing pressure
surface except for a region close to the leading edge, were distinguishing
features, as was wing taper. Wing motion apparently covered an arc of
approximately thirty degrees and this movement was similar to that proposed
for the first design.

    Following the design studies, the outcome of which bore little resemblance

----

[1] H.C.H. Townend. 'Note on the Possibility of Flight by Human Power'
    J.R.Ae.S. Vol. 41, p. 609, 1937.

to the type of aircraft most recently advocated by Elliott, he commenced an
ambitious and carefully planned systematic investigation into ways of obtaining
high lift and low drag using flapping flight.  It is hoped that this development
programme will culminate in the construction of the 'three-in-one' man-powered
aircraft.

   The suggested design is similar to a conventional glider in appearance but
is of modular construction, based on a common fuselage incorporating the single
pilot and controls and carrying a conventional tailplane and rudder, the latter
being provided with sufficient rigidity to support a pusher propeller for
operation in one of the three modes.  Power output is by rowing action, which
has been shown to be efficient because muscles of all limbs can be used, and
of the two man-powered configurations proposed, reciprocating motion can
conveniently be used for flapping and would be changed in the transmission to
rotational motion when the pusher propeller was used in conjunction with a
fixed wing.  (An alternative tractor location for the propeller was recently
being considered by Elliott), *(see Fig. 149).*

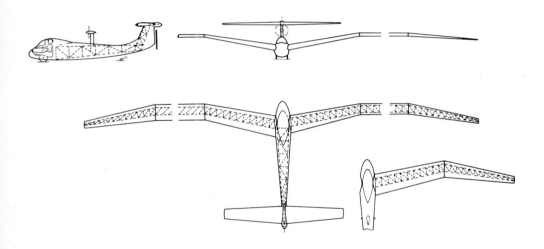

*Fig. 149*  Early layout of the 'fluttered' wing aircraft.

   A major objection to the adoption of the rowing motion in fixed wing man-
powered aircraft was based on the need to control in a single seat machine as
well as propel, and thus the hands needed to remain free.  The flutter motion
proposed by Elliott could well be best implemented using a rowing motion,
particularly as such movement may easily be adapted to provide the 'positive
aileron' propelling tip control suggested by Elliott as being one of the
advantages of this type of man-powered aircraft.  (This, it is claimed, will
overcome the control problem common to fixed wing aircraft when the inner wing
drag changes markedly during a bank).

   The operation in the fixed wing man-powered mode is regarded as a standby

arrangement, in view of the lack of knowledge of many aspects of the design
and performance of a fluttered wing system.  Fixed wing spars are a lightweight
tubular framework and each semi-span is divided into three segments.  Suggested
techniques for the control of roll in this mode include differential twist of
the wing root, but control features will not be finalised until studies of
lateral and longitudinal control in ground effect have been completed.

The fluttered 'swimmer' wing version, the propeller being dispensed with,
would have a similar articulated wing to that of the fixed wing man-powered
adaptation, with reduced chord and area being permitted owing to the lower drag.
A trailing edge flap was proposed, located on the inboard wing segments.

Use of a small outboard motor, located in a nacelle above the rear fuselage
and driving a similarly sited propeller of approximately 1 metre diameter,
defined the third mode of this versatile aircraft.  2.3 kW would be sufficient
to enable one to operate the aircraft with a reduced span, obtained by removing
the central segment from each semi-span.  Thus there would be more opportunities
for flying, as the aircraft would have an increased load factor and be able to
fly in less perfect weather conditions than a lightly stressed large span
machine.  The prime aim of the motor would be to allow easier investigation of
the handling characteristics, including the fluttered wing behaviour and its
influence on turns near the ground.

Because of the lack of data available concerning fluttered flight, its
influence on aerodynamic efficiency, structural integrity and the control
problems and advantages likely to accrue all need investigation.  An early
non-flying model of 1.525 metres span was used to illustrate the action of the
wing and root mechanism, a combination of cambered bending and twisting[1].
Other rigs operating or under construction in the early 1970s include a
simulator for the cockpit section, in which power output measurements of a man
in the rowing position may be made, and a lighter cockpit model incorporating
control systems and made after the form of construction proposed for the
prototype aircraft.  A 2.44 metre span flying scale model to test both fixed
and fluttered wing designs is reportedly under construction, as is a full-
scale test rig for aerodynamic investigations, to be carried out on a trolley
propelled along a runway, in a similar way to the measurements made on the
Southampton University aircraft wing profile.  Again both fixed and moving
wing modes will be tested, and comparative measurements of two-dimensional lift
to drag ratios made.

Structural development has been mainly directed at the wing.  Materials
considered for the framework included thin gauge aluminium tubes, balsa and
balsa-based carbon fibre.  The most critical parts of the structure will be
the outer wing sections, and resin impregnated carbon fibre sheet double-wound
into tubes is being considered for use here.

On a limited budget it is only possible to proceed so far.  Elliott has
found that in order to fully investigate detailed design features a more
precise form of experimentation and theoretical analysis is necessary.  It is
to be hoped that further work will be undertaken on this interesting and, as

---

[1] *In a brief note on his work, Elliott refers to similar suggestions for
obtaining lift portrayed in the drawings of Leonardo da Vinci.  Elliott's
contribution in a more detailed evaluation of this motion and the extension
to control remains unique.*

far as man-powered flight is concerned, unique interpretation of the fluttered wing form of propulsion. Its application in powered (particularly helicopter) flight could well provide the basis for the next stage of development.

## A CONTEMPORARY ORNITHOPTER

D.V. Curry, working in part at Bristol Polytechnic, has designed and constructed a man-powered ornithopter which lends support to some of the studies carried out by J.S. Elliott described above. The primary aim of the project was to model the propulsion system in such a way as to create wing movements analogous to natural flight. By introducing 'variable geometry'(of which the 'swing-wing' concept on fighter aircraft is one form this can take) the pilot could select a suitable wing form to give optimum performance during the various modes of flight. The design of the full-sized prototype began in 1960, the first successful propulsion tests being carried out in 1969. The results of the work were reported at the 1975 Man-Powered Aircraft Group Symposium in London[1].

As emphasised earlier, the identification of, and more particularly the reproduction of the motion of a birds' wings are critical factors. Curry has done this with reference to the three spatial axes shown in *Fig. 150 (a)*. Each wing consists of two segments, the inner segment corresponding to the humeral region of the bird's wing, the outer segment being analogous to the manus. Thus while the inner segment is restricted to up and down motion hinged about the centre line, the outer segment can move in two other ways independent of the motion of the inner segment.

To date, tests carried out on the prototype, the main parameters of which are listed in Table 18, have shown that the initiation of propulsion and acceleration is possible by means of the wing movements. In order to generate

Table 18. Curry's variable geometry ornithopter

| | |
|---|---|
| Wing span | 8.2 metres |
| Wing area | 9.8 m$^2$ |
| Aspect ratio | 6.9 |
| Weight of machine | 19.kg |
| Weight of undercarriage | 4 kg |
| Weight of pilot | 68 kg |
| All-up weight | 91 kg |
| Stroke plane angle | 76 degrees |
| Angle of attack at wing root | 14 degrees |
| Sweep angle* | 124 degrees |

\* defined for one downstroke - the aggregate
angle swept through by the leading edges of both wings.

[1] *D.V. Curry. 'The Propulsion of a Man-Powered Variable Geometry Ornithopter'. Proc. Man-Powered Aircraft Group Symposium, R.Ae.S., London 1975.*

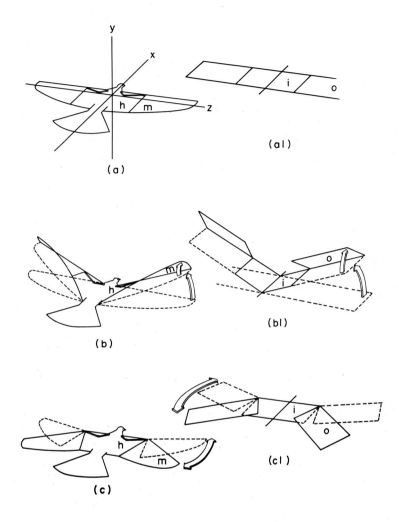

*Fig. 150* Wing positions in an idealisation of a bird,
with corresponding positions in a schematic of Curry's
variable geometry ornithopter. (See text for
identification of areas).

propulsion, it was found necessary to use large angles of twist of the outer
segments and if twisting was not used, no propulsion was forthcoming.

Although only limited trials have been reported to date, the attempt to
accurately simulate the initiation of natural flight is unusual among
proponents of ornithopters, and is of considerable interest.

*  *  *


    The danger in relying upon university research projects and other activities
where the effort of only one man is manifest in the design or construction of
an aircraft, however valuable the contribution might be technically, is that
they tend to operate in a vacuum.  It therefore becomes important that a body
such as the Royal Aeronautical Society is available to disseminate the results
of such work, or even to broadcast the fact that research into a particular
design problem or new concept is actually taking place.  The relationship
between the groups constructing full-scale aircraft, albeit of conventional
layout, and those who have carried out more specialised investigations into
unorthodox flying and propulsion techniques, or concentrated on improving small
but important items common to fixed wing and unconventional machines, does show
room for improvement, to the benefit of both camps.



*  *  *

*Fig. 148.* Model of the Salford University Aircraft

*16*

# One
# or
# Two
# Pilots?
## *— 'Toucan'*
## *and other Current Machines*

# 16

The evolution of man-powered aircraft design has become complicated by the fact that progress, even by the professional groups employing experienced aeronautical engineers, has in many instances been slow, and later designs which in current jargon might be described as second or third-generation machines, actually fly before less advanced machines being constructed on a long timescale basis. Obviously it is neither desirable nor practicable to continuously up-date one's design during such extended manufacturing runs to keep abreast of, say, new material developments, although in some cases a very simple modification might save the odd kilogram using the advantages of, for instance, carbon-fibre reinforced plastic resins. (See also Chapter 17).

We can accurately place the Southampton University aircraft (SUMPAC) and 'Puffin I' with regard to the Haessler-Villinger 'Mufli' and the Italian 'Pedaliante'. If we call the former two "second-generation" designs, 'Puffin II' is "third-generation". Difficulty arises when we try to place the Ottawa machine and Southend aircraft. The former was conceived over a decade ago, and is still under construction; the Southend aircraft can probably be placed somewhere between SUMPAC and 'Puffin II' on the ladder of technical achievement, although this is open to argument. The Japanese Linnet series can similarly be scattered at random within the same range - certainly these are not the most advanced designs and no significant performance improvement was obtained as a result of the numerous rebuilds, although the 'Egret' and 'Stork' derivatives represent definite steps forward.

As far as entry for the Kremer Competition is concerned, three United Kingdom contenders which, it was believed, would show significant advantages over other aircraft were the machines of the Weybridge Man-Powered Aircraft Group, the Hertfordshire Pedal Aeronauts and the RAF Team now at Cranwell. The Weybridge Group operating largely from within the British Aircraft Corporation, and under the leadership of P.K. Green, completed in 1971 a single seat mono-plane with a wing span of 37.51 metres. The Hertfordshire Pedal Aeronauts, revitalised from a nucleus of engineers employed at the now defunct Handley Page Aircraft Company, have flown their two seat aircraft, with a similarly enormous wing span, at Park Street (an ex-Handley Page airfield) near St. Albans The Weybridge and Hertfordshire aircraft, described below, must approach the current ultimate in wing span and lightness, particularly the Weybridge machine, being a single-seater[1].

Bearing in mind the performance necessary to complete a course of over 2 km, as required by the Kremer Competition rules, the Hertfordshire and Weybridge aircraft have taken the only path immediately open, i.e. increasing the wing

---

[1] *In the mid-1960s work by undergraduates at Glasgow University suggested that man-powered aircraft with very large wing spans would have very difficult lateral control characteristics.*

span, as a way of meeting these requirements.  Jupiter, based on the Hodgess-Roper aircraft described earlier and operated by the RAF team, has been the most successful, in spite of its lower wing-span.

However a complication has arisen in that not all man-powered aircraft groups now see the Kremer Competition as an immediate goal.  Led by Dr. Keith Sherwin at Liverpool University, who feels that the present state of the art does not justify an attempt on a figure-of-eight course, a growing number of enthusiasts are sympathetic to man-powered flight as a sport open to a much wider form of participation than that possible with aircraft such as those constructed at Weybridge and St. Albans.  Dr. Sherwin is the first to put this philosophy into practice in a serious manner, and in 1969 he took over the damaged airframe of 'Puffin II' and commenced modifications with the above aim in mind.

Before passing further comment, descriptions of these four machines are appropriate.

## THE WEYBRIDGE MAN-POWERED AIRCRAFT GROUP

The Weybridge Man-Powered Aircraft Group was formed in October 1967 around a nucleus of three employees of the British Aircraft Corporation (BAC). Objectives of the group were straightforward;  to design and build a man-powered aircraft which would have a better performance than existing designs. At that time the most successful machine was 'Puffin II'.  Towards the end of 1967 a preliminary design study was submitted to the Royal Aeronautical Society's Man-Powered Aircraft Group Committee and a grant of £50 was awarded, shortly to be followed by an identical sum given by the Weybridge Branch of the Society.

Main features of the aircraft, which has a low wing and a pusher propeller situated at the tail, are given in Table 19.

Table 19.  Main design parameters of the Weybridge aircraft

| | |
|---|---|
| Wing span | 37.51 metres |
| Wing area | 45 m$^2$ |
| Tailplane span | 4.19 metres |
| Tailplane area | 3.83 m$^2$ |
| Fin height | 3.45 metres |
| Fin area | 2.79 m$^2$ |
| Overall length | 5.63 metres |
| Propeller diameter | 2.13 metres |
| Wing section  -  Root | Wortmann  FX 68-M-180 |
|          -  Mid span | "        FX 68-M-160 |
|          -  Tip | "        FX 68-M-140 |
| Tailplane and fin sections | NACA 64-009 |
| Empty weight | 57 kg |
| All-up weight | 125 kg |
| Take-off speed | 7.3 m/s |
| Estimated cruise power | 0.23 - 0.30 kW |

Much of the work carried out during the first six months of the project was directed at manufacturing test specimens, particularly associated with the primary wing structure. From tests, it was decided to adopt a spar made using a box section of dural tubes for the booms and web members, bound together with glass fibre tape, a method similar to that proposed by Czerwinski in Canada during his investigations into structural optimisation. The wing sections chosen were designed for the Weybridge group by Dr. Wortmann of the Institut fur Aerodynamik und Gasdynamik in Stuttgart. Wortmann has given his name to a considerable number of aerofoils, the family being designed for operation at low Reynolds numbers, appropriate to those at which a man-powered aircraft operates.

By mid-1968 group active membership numbered one dozen, and considerable assistance in design and component manufacture was given by the BAC Training School. Overall design was complete and assembly was delayed by the fact that a suitable large area was not available at Weybridge airfield. Individual rib manufacture and the assembly of minor structural members could continue, however. In the fifth progress report issued by the group, pride of place was given to the announcement of a further grant of £200 given by the Royal Aeronautical Society, with a contingency grant of £100 provided by the local branch. Construction of a 1:10 scale wind-tunnel model was in progress. With this it was intended to check the flow pattern around the wing-fuselage junction and the rear fuselage. (Such local flows can cause increased drag if separation is not avoided, as occurred in the case of SUMPAC, although in this instance it was the propeller pylon which disturbed the flow at the trailing edge of the wing centre-section). By September, 1968 successful results had been obtained with this model, verifying that the overall layout was of a suitable form to inhibit excessive separation. At this point the group was forced to move its activities to Wisley airfield, a hangar being provided there for assembly of the machine. Unfortunately the move did not occur before a flood had destroyed the fin spar and damaged several ribs in the manufacturing shop at Weybridge.

The recent influx of money enabled the group to order a considerable quantity of materials, including aluminium alloy tubing to be used for the wing main spar. Several hundred feet of this were sent to BAC Filton to be chemically etched to a wall thickness of only 0.25 mm. This very thin-walled section was to be used in fabricating the outer wing spar units.

By mid-1969 progress on construction was well advanced, most of the inner wing section spar being complete, and the wing assembly jigs were being modified to include provision for location of the ribs on the main spar. Fuselage frame construction was well in hand, *(Fig. 151)* and many detail components had been completed. It was necessary at this stage to break into the £100 contingency grant, and a further £100 was given by the Royal Aeronautical Society in London. February 1970 saw the award of another £150 to enable completion of the construction; by this stage the propeller, of fixed pitch and with laminated construction, was ready for installation. The Weybridge designers concentrated on obtaining a reliable propeller section, rather than attempting to achieve a very high efficiency with the aid of a complex variable pitch arrangement.

Assembly of the major structural components was punctuated in March 1970 by a series of tests to verify the design integrity of the wing-to-fuselage joints and the wing primary structure. To carry out these tests, the wings were attached to the fuselage, the aircraft inverted, mounted on three aircraft jacks, and raised to a height sufficient to allow the wing tips to clear the ground.  Suitable weights were progressively placed on the wing spars until a loading was reached identical to that likely to be met under 1.5 g accelerations of the aircraft during flight.  Measurements were made of the tip deflection, and the wing incidence control mechanism was checked for freedom of movement at the various load values, *(Fig. 152)*.

The exercise above was particularly necessary when it is noted that to effect control of the Weybridge aircraft, the complete wing was pivoted about joints under the pilot's seat.  Each wing could of course be pivoted independently.  During loading tests it was discovered that above an equivalent load of 0.5 g the pivot mechanism fouled the main fuselage structure, and modifications were needed to overcome this problem.  A feature which could greatly affect the aircraft performance was also found during these experiments; near the wing root considerable shear deflection of the spar caused some ribs to distort, and it was suggested that locally some ribs unattached to the spar could be used to avoid these deformations, which could change the local lift and drag characteristics beyond recognition.  An idea of the wing tip deflections likely to be encountered in severe manoeuvres with man powered aircraft is obtained from the fact that under static conditions the wing tip is 0.1 metres above the level of the wheel axle, but under 1 g acceleration the height of the tip above the same datum reaches 3.5 metres.  It was concluded from the static load tests that the wing primary structure was sound enough to meet the design load conditions, but further modifications to the pivot mechanism were needed following later trials.

Many minor tasks remained to be done, including fitting of the pilot's seat and harness, the pilot operating the pedals from a reclining position, *(see Fig. 153)*.  The bevel gear box and transmission shaft to the rear-mounted propeller were fitted within the fuselage frame, and control cable runs were installed.

On Saturday, 5th December, 1970, the aircraft, by now named 'Dumbo' was taken out of its hangar at Wisley and ground taxi runs were made on the apron.  No special problems were encountered, but a note was made of the difficulty in handling such a fragile craft.  The aircraft was then placed in a position ready to fly as soon as weather permitted.  It was then found necessary to move the machine back to its birthplace at Weybridge, and it was rolled onto the runway at this airfield for the first time on 27th February, 1971.  Wind speed was in excess of 5 km/h, and it was therefore not considered safe to attempt a flight.  Short ground runs were made however, and by far the most important lesson learned at the time was the lack of knowledge of the power output of the pilot.  Construction of a rig was commenced on which quantitative measurements of power output could be made under representative conditions, but it was intended that the flying programme should continue in parallel with rig tests.

Between the beginning of March and the 5th April, 1971, four further series of taxi runs were carried out to familiarise the crew with handling character-istics.  On the last trial a take-off attempt was planned, but had to be abandoned due to a buckled front wheel and gearbox malfunction.  The pilot's assessment of the handling characteristics was generally encouraging.  Response in roll was good, feeling the same as that of an aircraft fitted with conventional ailerons.  Within the limitations of ground trials, indications were that yaw control might be improved with larger permitted rudder movements.  The group were optimistically able to report at the end of this series of trials that they possessed a man-powered aircraft capable of flying once minor 'teething troubles' had been sorted out.

Unfortunately, before a flight could be successfully attempted, the aircraft was damaged by a gust of wind, repairs being necessary to the wing, tailplane and rudder. Balsa members were broken in the accident, but the alloy tube primary structure remained undamaged.

By August 1971, repairs had been completed, and taxiing trials were resumed. The following extract from a Weybridge Group Progress Report details the first flight, *(see Fig. 154)*.

"On Saturday the 18th September, at about 10.00 a.m., the Weybridge Man-Powered Aircraft, piloted by Chris Lovell of the Surrey Gliding Club, achieved its first take-off. Two short flights were made on the runway at Weybridge, the second of which covered about 50 yards (45 metres) at a height of about 3 feet (1 metre). This is a milestone in the development of the aircraft and should increase the confidence of our supporters."

Although take-off was achieved at the first attempt on Saturday the 18th, many runs were made on the previous Monday evening. During the evening session it became obvious that wing incidence was the key to a successful take-off. Incidence was progressively increased and finally set at 3.3 degrees.

The following information relates to these flights:

| | | |
|---|---|---|
| Wind speed | – | zero |
| Weather | – | warm, bright sunshine |
| Take-off speed | – | 15 miles/h (24 km/h) |
| Take-off run | – | 100 yds (approx) (91 metres) |
| Pilot's comments | – | power required as well within his capacity |
| | – | adequate roll and pitch control, negligible yaw control |
| | – | favourable overall impression |

Further flights of the Weybridge aircraft were made, but it has now been moved to Cranwell, where Sq. Ldr. John Potter's team took over development, renaming the machine 'Mercury'.

## THE HERTFORDSHIRE PEDALNAUTS

Another serious contender for the Kremer prize is the 'Toucan', the creation of the Hertfordshire Pedal Aeronauts (HPA), shown in *Fig. 155*. Of marginally greater wing span than the Weybridge aircraft, the Toucan is characterised by its two seat tandem crew arrangement and a large slab-sided fuselage. Data on this aircraft is given in Table 20.

The group was founded in September 1965, and gradually built up around a small production facility provided by the Handley Page Aircraft Company at Radlett airfield. Early activities centred on feasibility studies, the aircraft being aimed directly at the Kremer Competition, and one feature which stood out during this period, and later in the more detailed design study, was the significance of the solution of the many practical problems, as opposed to the comparatively well defined theoretical approach to the design of such a machine. By December 1966 the study was completed and was submitted to the Royal Aeronautical Society with a view to obtaining a grant to assist in construction. As a result, an initial sum of £500 was given to the HPA.

Table 20.  Main features of the HPA 'Toucan'

| | |
|---|---|
| Wing span | 37.5 m (later increased to 43.5 m) |
| Wing area | 55.8 m$^2$ |
| Wing section | NACA 63$_3$ 618 |
| Aspect ratio | 25 |
| Wing weight | 36.4 kg |
| Empty weight | 66 kg |
| All-up weight | 202 kg |
| Wing loading | 3.61 kg/m$^2$ |
| Cruise speed | 7.62 m/s |

    With the study restricted to fixed wing aircraft, it was concluded that no
marked improvements could be anticipated if a single seat aircraft was built,
the group having initially only SUMPAC and 'Puffin I' as the yardstick for
single-seat capabilities.  Advantages of the two seat configuration chosen, in
which both crew pedalled, the aft member in addition controlling the aircraft,
were claimed to be a higher power to weight ratio and the prospect of better
continuity of power output, particularly during difficult manoeuvres such as
turns.  In adopting a mid-fuselage wing position, the group aimed to make
improved utilisation of ground effect.  The short fuselage, it was felt, would
negate the use of a conventional rudder, as the turning moment would be too low.
Lateral control is therefore achieved using slot tip ailerons and wing tip
spoilers.  Claimed attraction of these slot tip ailerons is that response time
is considerably reduced when compared with conventional ailerons, a very
important point when viewing the flights of high aspect ratio man-powered
aircraft in general.  An all-moving tailplane is retained, (see Fig. 156).
One benefit derived from the use of a fixed rudder is that the forces produced
by rudder movement and which have to be transmitted through the rear fuselage
are almost eliminated.  As a result this portion of the fuselage may be
considerably lightened;  this was one of the reasons for the concentration of

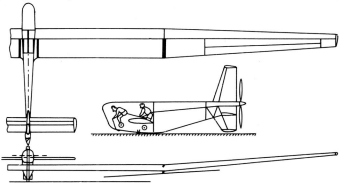

Fig. 155  The Hertfordshire Pedalnauts' 'Toucan'

control surfaces on the main wing.  Employment of a pusher propeller behind the tail would also have suffered if a moveable rudder had been chosen, as the flow distribution through the propeller disc would have changed with every movement of the rudder.  Like the Weybridge aircraft, wing dihedral is large, tip deflections of over 2 metres occurring during 1 g flight.  The designers propose to use this to advantage as a stabilising force in automatically cutting out slide-slip.  Wing deflection is kept within what is considered an aerodynamically acceptable limit by using a relatively thick aerofoil section, the laminar flow NACA 63$_3$ 618, with a thickness to chord ratio of 0.18, *(see Fig. 157)*.

Transmission from the pedals runs via light weight chains to the ground wheel, which is driven at all times, and thence to a torque tube connected to the propeller.  During the ground run the aircraft is intended to assume a nose-down attitude with a small nose-wheel offering support, until a speed of approximately 9 m/s is achieved, at which take-off and climb to cruise altitude is deemed to occur.  A cruise speed of 7.6 m/s allows for operation at a reduced power level over an extended period.

Fairly conventional structural techniques and materials have been chosen for much of the primary and secondary wing structure, although the primary fuselage frame supporting the pilots is compact compared with the angular frameworks incorporated in many of the earlier aircraft, being a box-section light alloy unit of 30 gauge sheet joined by riveting, *(see Fig. 158)*.  Spruce is used extensively, the I - section spar being the main claimant to this material.  A strong leading edge section is also featured, but in many of the non-load bearing parts of the structure, where preservation of a reasonable curvature is the sole design criterion, materials of the lowest possible density are used, including very light grades of balsa and, more commonly, expanded polystyrene.  The latter sheets the leading edge region of the wing, but most of the structure is finally covered with 'Melinex'.

Target dates for the first flight of 'Toucan' were put back several times, not least because of problems brought about by the failure of the Handley Page Company, with the resultant loss of jobs by many of the team, under the leadership of Martyn Pressnell, who has the official title of the Chairman.  The relocation of many of the members within the North London area assisted in ensuring that personnel problems were kept to a minimum, but the closure of the airfield was a second blow.  Construction *(Fig. 159)* proceeded at another ex-Handley Page field, Park Street, but it was this lack of facilities for construction and flying of 'Toucan' and other man-powered aircraft, including the Southend and Woodford machines, that prompted the HPA to propose to the R.Ae.S. Man-Powered Aircraft Group that:  "The Royal Aeronautical Society Man-Powered Aircraft (Group) Committee endeavour to establish a permanent storage and flight test facility for man-powered aircraft, as a matter of urgency."  It was suggested that Cranfield might be the most suitable location for such a centre, but such a proposal was to many totally impractical, man-powered aircraft groups being scattered throughout the country and unable to travel at a moment's notice should a rare day on which ideal weather conditions for flight occur.  Man-powered flight is sufficiently attractive to the mass media to cause any publicity-minded organisation to think twice before turning down requests for goods or services, as several groups have found, and for which they are extremely grateful.

Probably the clearest indication of the potential of 'Toucan' was given by Pressnell during a lecture in London in May 1966, at which he estimated that the power requirement per man for flying this aircraft was 0.15 kW, a figure within the limits of our 'average fit man', and which would apparently offer flights of unlimited duration to an athlete.

The first flights of 'Toucan' were made in December 1972, and the longest flight made to date is 640 metres. The wing span of the aircraft has been subsequently increased to 43.5 metres, and further flights have been carried out, although several culminated in structural failure of one form or another. During 1976 repairs have been carried out, and the aircraft should fly in 1977.

## KEITH SHERWIN'S 'LIVERPUFFIN'

The 'odd-man-out' among the current aircraft in the United Kingdom described here is Dr. Keith Sherwin's 'Liverpuffin'. Hardly recognisable as a rebuild of 'Puffin II', 'Liverpuffin' is not directed at the requirements of the Kremer Competition, having a limited range capability, but rather is aimed at the idea of man-powered flight as a sport. Table 21 gives a brief description and comparison.

Table 21.  Comparison of some design parameters of 'Liverpuffin' and 'Puffin II'.

|  | 'Liverpuffin' | 'Puffin II' |
|---|---|---|
| Wing span | 20.1 metres | (28.3 metres) |
| Wing area | 29.7 m$^2$ | (36.2 m$^2$) |
| Tailplane span | 3.35 metres |  |
| Wing section | Wortmann FX-63137 |  |
| Wing weight | 33.6 kg | (29.5 kg) |
| Empty weight * | 57.7 kg | (52.6 kg) |
| All-up weight | 127 kg | (116.1 kg) |
| Range | 100-250 metres | (2.5 km) |
| Altitude | 1-1.5 metres | (3 metres plus) |

* A contingency allowance of 9 kg is excluded from these figures.


'Puffin II' arrived at Liverpool University in a comparatively fair condition, and the group there were able to use many of the components in redesigning the layout. Parts retained include the primary wing structure, with a reduction in span (the 'Liverpuffin' uses the same Wortmann laminar flow aerofoil section as 'Puffin II'), the propeller, tailplane, fin and rudder parts, and the pilot support frame. The fuselage was changed to a pod and tail boom combination similar to that proposed by Lippisch in his post-war proposal for a man-powered aircraft with similar aims to that of Sherwin's machine. Such a pod minimised transmission distances, allowing use of a simple belt to drive the ground wheel, from which a shaft ran to the propeller at the rear of the pod. A high wing was retained, and the tail boom projected from the base of the pod, the propeller being sufficiently high to clear the boom.

Extensive use of expanded polystyrene, most evident in the wing structure, *(Fig. 160)* distinguishes the machine from many of its contemporaries.  By using a hot wire cutting tool one is able to shape large polystyrene components with comparative ease, and excess material incidental to that required to bear any load may be removed in the form of lightening holes or by hollowing out of the component.  Much of the wing leading and trailing edge of 'Liverpuffin' is formed using expanded polystyrene.

Sherwin's view that man-powered flight should begin unambitiously as a sport and proceed, as the gliding movement did, to more serious flying once sufficient experience has been gained, is attractive in that it would open up the prospect of participation in this form of motivation to a wider group of people than those within the rather specialised university and industrial teams which have been responsible for most advancement to date.  Critics would claim that even these extremely able design teams with first-rate manufacturing expertise and materials to hand have so far been unable to construct and fly an aircraft which is capable of repetitive flights of predictable length, course and altitude, albeit less than the requirements of the Kremer Competition.  It must be acknowledged that materials technology is available to a much wider public, and is progressing at a considerable rate.  However, aerodynamics of very low speed flight is a science which has to some extent built up around the unique requirements of man-powered aircraft, as is only too clear from the lack of data on aerofoil characteristics over Reynolds number ranges relevant to these machines, and improvements are almost certain to come from aerodynamicists in research and development establishments.  Implementation of these improvements will not be easy.

'Liverpuffin' is an integral part of the undergraduate course in engineering design at Liverpool, and is classed as a major design project.  Its choice in this rôle was made for three reasons - the complexity of the project is of a similar order to that found in industry;  practical work is comparatively simple and readily carried out on available machinery, in spite of complex theoretical analyses associated with man-powered flight;  and there is no standard solution to the design problems, thus the exercise becomes a good test of initiative.

Apart from the performance aims listed earlier, it was within the students' terms of reference to produce an aircraft which was robust, easy to construct, transportable, and generally manageable both when in the air and, possibly one of the most important criteria, when on the ground.

Even 'Liverpuffin' is probably too large and sophisticated to meet all Sherwin's requirements as listed above.  He predicts man-powered aircraft costing up to £500, fabricated in a number of sections, each typically of 9 metres length or less.  Spans of the order of 14 metres would allow very limited flights and provided that the courses were not too ambitious, a competitive sport could be built around these machines.  It would, in view of the fact that no major man-powered flight competition has yet been won[1], be preferable to fix the type of course over which competitions could be run following flights of at least one or two of the aircraft which are judged typical of the class likely to participate in such events.

In order to become sufficiently attractive, a sporting man-powered aircraft would have to be able to fly in circumstances more severe than the still air

---

[1] *The Peugeot prize is not included in this category.*

conditions required by 'Puffin' and the present generation of potential Kremer Competition contestants.  This is fully appreciated by the Liverpool group, and is manifest in their prototype as an increase in empty weight compared with the considerably larger 'Puffin II' from which it was derived.  Sherwin foresees the use of the many disused airfields throughout Britain as flying grounds for these aircraft, and club/national competitions with a direct incentive to improved aircraft design and development.  As the aircraft would be designed such that they could be flown by average cyclists, athletes could obtain extended flight durations and the more robust craft could possibly be sufficiently stressed to make use of low-altitude convection currents.

The two camps, those pursuing the Kremer prize and the less ambitious but equally enthusiastic sporting group, represented by Dr. Sherwin but receiving more support as time progresses, are not in direct conflict.  Each faction is likely to learn much from the other, and it is probable that both will progress in parallel for some years.  Emphasis will undoubtedly shift towards the 'Liverpuffin' philosophy if 'Toucan', 'Jupiter' and the Weybridge aircraft, having now joined 'Jupiter' and renamed 'Mercury' (see below), fail to satisfy the Kremer Competition requirements, and it is unlikely that projects of the same size, both physically and on an organisational scale as these latter two will be seen again in Britain for some years.

'Liverpuffin' has met the performance aims set by Dr. Sherwin and his team.  The secondary wing structure, fabricated in expanded polystyrene was found to be satisfactory, and the first flight, made on 18th March 1972, showed that control with 'rudder only' was possible close to the ground.  The flight was made at Woodvale airfield.

## 'JUPITER'

The most successful aircraft, and the current holder of the distance record (1.07 km, achieved on 29th June 1972), is the 'Jupiter'.

This aircraft, based on the Hodgess-Roper machine, was constructed at Halton RAF College, and involved 4000 man-hours.  Some 100 individuals were involved in the construction, including many apprentices.  It is a single-seat aircraft and the main data is given in Table 22.

Table 22.  Technical data on the 'Jupiter' aircraft

| | |
|---|---|
| Wing span | 24.38 metres |
| Wing area | 27.9 m$^2$ |
| Tail area | 2.41 m$^2$ |
| Wing section | NACA 65$_3$618 |
| Aspect ratio | 20.7 |
| Overall length | 8.99 metres |
| Empty weight | 66 kg |
| All-up weight | 136 kg |
| Cruise speed | 17.5 kt |
| Stalling speed | 16.0 kt |

The Jupiter is shown without its cockpit fairing in *Fig. 161* and in flight in *Figs. 162 and 163*.  While flights are expected to continue in 1977 much has already been learned concerning the major factors influencing good design and the handling characteristics.

With regard to design and assembly, the group emphasise the importance of keeping the structural weight to a minimum and recommend that to achieve this the full strength of materials (not just the conservative published strength values) should be fully exploited.  It was also stressed that the use of adhesive should be closely controlled, as these could incur a substantial weight penalty.

Concerning flight trials, experience showed that special training was required to put the pilot into peak physical condition for sustained flights. The wind was the most influential factor in determining range and take-off distance.  In a 5 knot head-wind take-off was comparatively easy to achieve. If a tail-wind of the same speed was present, take-off was virtually impossible.

The most significant conclusion was that ground effect apparently gave little advantage, the pilot always feeling more comfortable at 6 metres altitude than at 1 metre.  This is contrary to the argument put forward by many workers in the field, including the latest proposals of Rear Admiral Goodhart, discussed later in this chapter.

The 'Jupiter', including design and construction at Woodford, and rebuilding at Halton, has been developed over a period of twelve years.

The culmination of this work has been the achievement of the official longest flight by a man-powered aircraft with no assisted take-off, this being 1.07 km, the flight taking place at RAF Benson on 29th June, 1972.  The longest flight made (not officially observed) was 1.23 km on 16th June in the same year.  These followed the first flight on 9th February 1972.  In all cases the pilot was Sq. Ldr. Potter.

The raising of the prize, now £50 000 (see Appendix V) for successful completion of the Kremer figure-of-eight course, has increased interest considerably.

## THE MOST RECENT KREMER COMPETITION CONTENDERS

A number of projects are under-way in Europe and other parts of the world, which should provide considerable interest over the coming years.  Types of machine under construction include fixed wing, ornithopter and autogiro aircraft.

M. Hurel, noted for his high aspect ratio wing powered aircraft of the 1950s has constructed a parasol tractor monoplane having a span of 42 metres and an aspect ratio of 30.  This aircraft, illustrated in *Fig. 164*, is being flown at Le Bourget Airport by M. Thierard, a competition cyclist and pilot.

With an empty weight of only 66 kg, and a wing loading of 2.5 kg/m$^2$, which is only about 50 per cent of that of the nearest comparably sized aircraft, 'Toucan' (albeit a two-seater), Hurel's machine should receive increasing attention when the next 'flying season' approaches.

Other aircraft recently completed include a monoplane in Vienna and a parasol type wing machine in Brussels.

Interest in the United States was until recently (see later) centred
on MIT and the University of Oklahoma, where two-seat canard biplanes have
been built.

The Massachusetts Institute of Technology 'BURD' uses a biplane configuration
to enable the wing span to be reduced facilitating turns and to minimise weight.
The main parameters of 'BURD' are given in Table 23.

Table 23.  Specification of the MIT 'BURD'

| | |
|---|---|
| Wing span | 18.9 metres |
| Wing area | 59.4 m$^2$ |
| Canard span | 6.4 metres |
| Canard area | 5.6 m$^2$ |
| Length | 8.2 metres |
| Propeller diameter | 3.0 metres |
| Empty weight | 58.1 kg |
| All-up weight | 181.6 kg |
| Cruise speed | 28 km/h |

A pusher propeller is located behind the vertical stabiliser, visible in
*Fig. 165* and in the arrangement drawing, *Fig. 166*, and the canard surface is
mounted on the front of the fuselage, which is largely made of aluminium tubing.
The wings and other flying surfaces are constructed using balsa wood, with
plastic film coverings.

The crew are mounted in tandem in the conventional cycling position:  a
flexible cable chain of new design is used to transfer rotational movement from
the pedals to the propeller shaft.  In this chain the conventional links and
rollers are replaced by two cables joined by moulded connecting discs of plastic
spaced at the correct pitch.  The cable sides made the chain flexible in all
directions, and the plastic discs are self-lubricating.  This enables a
90 degree change of direction to be made without difficulty.  The total
transmission system weighs 1.6 kg.

Taxi trials highlighted areas which needed further development.  The most
significant was the canard surface area.  In addition, steering capability was
added to the front wheel and the transmission chain is the subject of continued
development.

Although the first flight attempt culminated in an almost complete structural
collapse in 1975, it will be interesting to see how this, the most sophisticated
man-powered biplane, performs when rebuilt.  The reduction in span possible,
indicated by the author's research in 1966, may well be the main feature in
improving handling in the turn, a critical manoeuvre in the Kremer figure-of-
eight course which 'BURD' is designed to attempt.

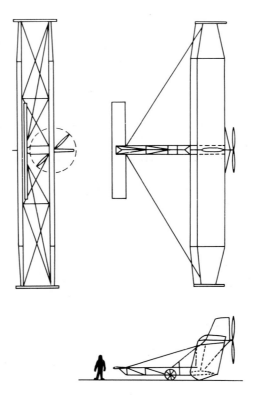

*Fig. 166*  General arrangement drawing of the MIT 'BURD'

## SUCCESS IN THE UNITED STATES

The year 1976 saw the first officially observed man-powered flight in the
United States.  The aircraft involved, the Olympian ZB-1 monoplane, was designed,
constructed and flown by Joseph A. Zinno, of Centredale, Rhode Island.

Joseph Zinno's efforts in designing and constructing his aircraft, a task
which took four and a half years, involving 7000 man hours, and the success of
the machine, will give encouragement to the individual enthusiasts who may
believe that only the resources of universities of the aircraft industry can
provide the technology needed to produce a successful man-powered aircraft.
Zinno commenced the design of the Olympian ZB-1 in 1972, and the machine
received a Federal Aviation Administration (FAA) Airworthiness Certificate on
19th January, 1976.  Given the registration number N1ZB, this is the first
man-powered aircraft in the United States to be given an airworthiness
certificate.

The Olympian ZB-1, illustrated in *Fig. 167*, is a single seat monoplane of
conventional layout, with a pusher propeller mounted behind the cockpit.  With

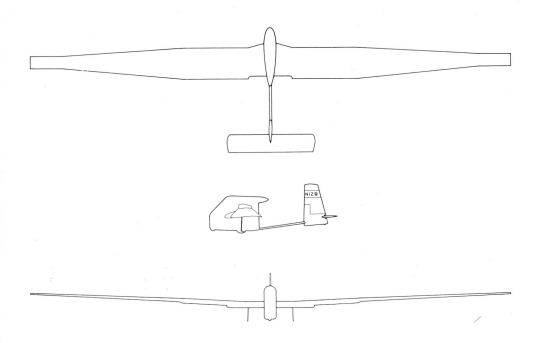

*Fig. 167*  A three-view drawing of the Olympian ZB-1, which is
the first man-powered aircraft in the United States to be given
an FAA Airworthiness Certificate

a wing span of 24 metres and an empty weight of 67 kg, the Olympian ZB-1 is
similar in size to the 'Jupiter' machine.  Further comparison of the data
given in Table 24, with that in Table 22, indicates many other areas of
similarity.

The first flight of the Olympian ZB-1 took place at Quonset Point Airport,
North Kingstown on 21st April, 1976.  The official observer recorded this
flight as lasting five seconds at an altitude of 0.3 to 0.45 metres.  The
estimates of the distance covered range between 10 and 30 metres, an
indicated airspeed of 27 km/h being achieved.

Structurally, the Olympia ZB-1 uses conventional assembly techniques
common to a number of other man-powered aircraft.  Spruce, birch plywood and
balsa are extensively used in the flying surfaces and the cockpit fairing,
while an aluminium and steel primary structure supports the pilot and
transmission system.  As in Dr. Sherwin's 'Liverpuffin', a polystyrene type
foam is used for rigidity and profile maintenance in the region of the wing
leading edge, extending back to cover about 35 per cent of the wing surface.
The structure was covered using 'Mylar' film (similar to 'Melinex').

Table 24.  Technical details of the Olympian ZB-1

| | |
|---|---|
| Wing span | 24 metres |
| Wing area | 29 m$^2$ |
| Aspect ratio | 19.5 |
| Tailplane area | 4 m$^2$ |
| Rudder area | 1.2 m$^2$ |
| Propeller diameter | 2.65 metres |
| Aerofoil section (inboard) | FX MS 72 150B |
| Aerofoil section (outboard) | FX 63 137 |
| Empty weight | 67 kg |
| All-up weight | 132 kg |
| Stall speed | 24 km/h |
| Take-off speed | 27 km/h |
| Cruise speed | 32 km/h |
| Sink speed | 0.26 m/s |

The feature which readily identifies the Olympian ZB-1 from most of its contemporaries, is the single boom, in the form of a 10 cm diameter aluminium tube. One of the primary advantages of a boom-type fuselage is its low weight, while permitting one to keep a comparatively long moment arm for effective operation of the tailplane and rudder. There can, however, be problems associated with the design. The selection of a short section for containing the pilot can lead to a disproportionately high drag, due to flow separation near the trailing edge of the section.

## THE 'DRAGONFLY'

One fixed wing man-powered aircraft currently being designed in the United Kingdom does, in the view of the author, go some way towards meeting the requirements cited in Chapter 17 by Haessler. This machine, the 'Dragonfly', is based on the wide experience, both as a pilot of 'Jupiter' and in design and construction aspects, of R.J. Hardy, who is based at Scottish Aviation, Prestwick[1].

In arguing that the best aircraft from the practical point of view, were SUMPAC, 'Puffin' and 'Jupiter', Hardy opts for the lower wing spans which are common characteristics of these early machines, and, certainly in the case of SUMPAC, the comparatively low number of man-hours needed for construction. As discussed elsewhere in this book, one of the most serious flying characteristics of man-powered aircraft, particularly those with very large wing spans (in excess of 30 metres) is their lack of ability to turn without rolling over due to wing tip stalls. The combination of a comparatively low span (24.4 metres) and the optimisation of the design so that it can successfully operate outside (or near the limits of) ground effect form the basis of the 'Dragonfly' design philosophy. The additional altitude can be put to good use in attempts at the

---

[1] *R.J. Hardy. The optimised man-powered aircraft. Proc. Man-Powered Aircraft Group Symposium, R.Ae.S., London, February 1975.*

Kremer figure-of-eight course, permitting greater room for manoeuvre at heights of up to 20 metres.  A general arrangement of 'Dragonfly' is given in *Fig. 168*.

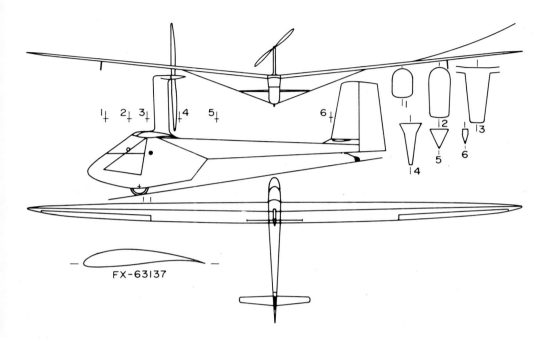

*Fig. 168*  General arrangement of the 'Dragonfly'.

## REAR ADMIRAL GOODHART'S 'NEWBURY MANFLIER'

Another UK contender for the Kremer competitions with a good chance of attempting one of the courses during 1977, is the 'Newbury Manflier'.  This aircraft is the practical result of ideas first put forward by Rear Admiral Goodhart in 1974[1].  The basis of Goodhart's philosophy is the better use of 'ground effect', or the drag benefits which are to be gained by flying close to the ground.  Although some modifications have been made to the design since the original proposal was put forward, this being illustrated in *Fig. 169*, the concept of two crew being located in 'pods' on the wing remains, although these have now been moved inboard from the wing tips.

Initially aiming for a wing span of 55 metres, with an all-up weight of 321 kg, for which the theoretical power requirement per man was only 0.13 kW, the latest layout, incorporated in a machine which is scheduled to fly in early 1977, has a wing span of 42.5 metres and an all-up weight of 209 kg.  The

[1] *H.C.N. Goodhart.  A man-powered aircraft with power to spare.  Aeronautical Journal, Vol. 78, No. 765, September 1974.*

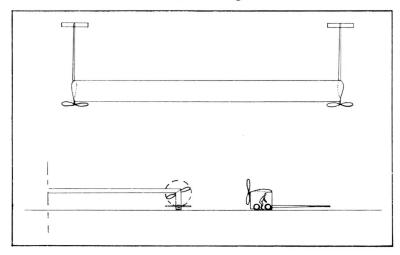

*Fig. 169* Rear Admiral Goodhart's proposed design, which
endeavours to make better use of 'ground effect'.

distance between the pilot 'pods' has been reduced to approximately 20 metres,
the overall aspect ratio being high at 29.  It is to be hoped that this machine
does not suffer the turning problems associated with some of the other large
span aircraft.

* * *

While Goodhart's aircraft is large by any standards, its wing span is
exceeded by a small margin by the aircraft proposed by students at Nova Scotia
Technical College.  With a span of 56.5 metres, and a crew of seven, located
in wing pods and in the conventional fuselage, this ambitious project will be
of some interest.  A small (2 metre span) model has been flown, and it is to
be hoped that the design will shortly move to the hardware stage.

* * *

*Fig. 151.* View of the Weybridge aircraft, prior to covering,
showing the primary and secondary fuselage frames.

*Fig. 153.* Cockpit of the Weybridge machine.

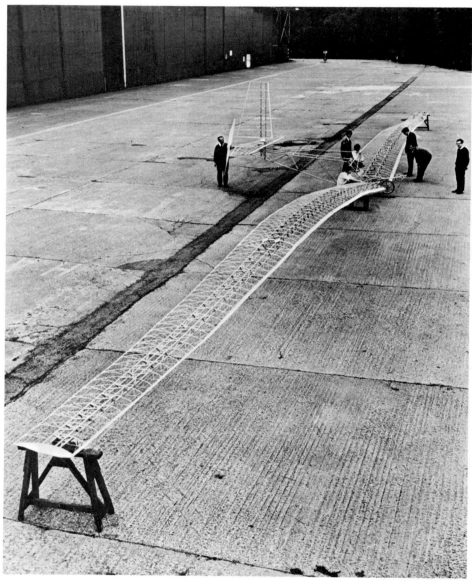

*Fig. 152.* The Weybridge aircraft photographed before
the fuselage profile was added.

*Fig. 154.* Weybridge aircraft in flight.

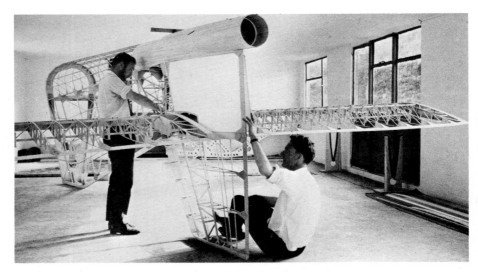

*Fig. 156.* The rear fuselage and tail unit of 'Toucan'.
Note the large all-moving tailplane.

*Fig. 157.* Details of the wing structure of 'Toucan'
are visible in this photograph. The nose
of the aerofoil had yet to be added.

*Fig. 158.* The riveted frame for supporting the two crew and primary drive mechanism of 'Toucan'.

*Fig. 159.* This view of 'Toucan' shows to advantage the large wing-span and built-in dihedral.

*Fig. 160.* Wing structure adopted by Sherwin for 'Liverpuffin'.
The main spar of 'Puffin II' is retained.

*Fig. 161.* Jupiter without cockpit fairing.

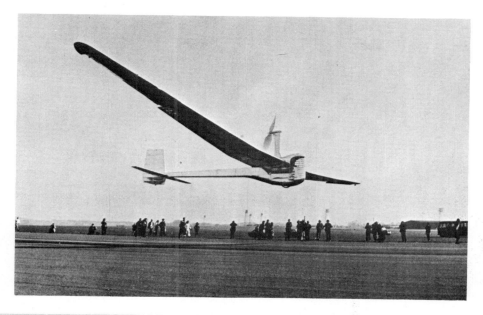

*Fig. 162*   Royal Air Force 'Jupiter' man-powered.
*& 163.*     aircraft in flight, March 1972.

*Fig. 164.* The Parasol Aircraft of M. Hurel

*Fig. 165.* The M.I.T. BURD Biplane.

# 17

# Future
# Prospects

# *17*

The maturing process of man-powered flight has, after a transitional phase, begun to follow two separate but largely complementary paths. It is noticeable that the initiative in both instances has come from within Britain, and the scale of the almost world-wide reaction during the 1960s and early 1970s to the movement given the initial impetus by the founders of MAPAC, and perpetuated by the Royal Aeronautical Society, is evident from the description given in earlier Chapters of the activities in various countries.

Purists who consider that man-powered flight as such is only valid when accompanied by an unaided take-off have probably had the most influence on our thinking over the past two decades. While in some ways this has throttled development, the technical expertise and knowledge gained as a result of groups striving to achieve long unaided flights has put us in a much more advanced state as far as the ability to design and construct successful man-powered aircraft is concerned, than would have been the case had no incentives been provided.

The future of man-powered flight has been the topic of numerous closing remarks contained in papers written by those closely associated with its development. In 1961 Helmut Haessler, in a communication to B.S. Shenstone, included notes on a paper entitled 'Soaring and Man-Powered Flight'. In it Haessler considered the possibility of flights with assisted take-off to altitudes of 20 to 30 metres and performance which could be attractive to an average non-cycling pilot. Forms of assisted launch included rubber bungee, car tow and a winch. Release height would ideally be at an altitude at which the chance of finding a thermal updraft equal to the sinking speed of the aircraft is of the order of 20 per cent; by using man-power the probability of retaining or increasing altitude would, Haessler considered, be ideally around 80 per cent. Thus while a man-powered flight under such conditions would not be a certainty, its occurrence would be at a significantly higher frequency than a standard glide under identical atmospheric conditions. Haessler compared this form of flying activity with glider performance, particularly the frequency of flights possible. In his experience, winching of gliders, where release height was approximately 150 metres, took place at a frequency of about three per hour, with each flight lasting of the order of 5 minutes. Thus preparation time in this instance is three times the flight duration. If it was possible to reduce the release height using man-powered aircraft, Haessler argued that the preparation time for each aircraft could be reduced to about 5 minutes. With four take-offs per hour, the flying time could be 10 minutes per launch, representing a 100 per cent increase in time in the air.

The type of man-powered aircraft suggested as being suitable for this form of flying would have a span of 20 - 22 metres, assuming a single seat monoplane, somewhat smaller than that needed for unassisted take-off. Haessler felt that a two seat machine having a wing span of 25 metres would be appropriate as a training aircraft.

This category of man-powered flight has yet to be practically investigated. The major drawback might be the structural weight necessary to meet the strength requirements of the more arduous flight envelope of this machine, which is likely to approach the extremities of that corresponding to a simple glider. If this proves to be the case, the ratio of the probability of flight based on thermal strength or man-power of 1:4 is likely to be reversed, the thermal almost achieving the importance which it maintains in pure gliding.

Writing in the *Journal of the Cambridge University Engineering Society* in 1965, Shenstone commented on the state of the art of man-powered flight and made some personal observations on future trends. He emphasised the fact that man-powered flight as such was a marginal occupation, and was unlikely to become a general sport. Cost was cited by Shenstone as another drawback to its acceptance as a sport; as much of the labour in construction of these aircraft was given free, their cost was difficult to assess. However, he estimated that one would have to pay approximately £100 per pound weight for a man-powered aircraft. At that time a Hawker Siddeley Trident aircraft cost in the order of £40 per pound of structural weight.

Shenstone foresaw the possibility of races between man-powered aircraft. The FAI had provided the basis for world records for man-powered aircraft and these would be worth competitive effort. In the paper he returned to his earlier point concerning the uncertainty of flight, stating that an absolute duration limit may be reached, because of man's power and other difficulties.

At a joint meeting of the Canadian Aeronautics and Space Institute and the American Institute of Aeronautics and Astronautics, held in Toronto in 1970, Czerwinski took the opportunity to give us his predictions concerning man-powered flight as a leisure activity and an educational aid. His association with the Ottawa man-powered aircraft project, discussed in detail in an earlier chapter, has been closely associated with structural design, and he naturally emphasised weight reduction as a major contributory factor to improvements in aircraft design and performance. Czerwinski went on to state:

"In the case of man-powered aircraft the engineering challenge is very great as the domain of physical and technical sciences involved is not sufficiently covered by the available state-of-the-art. The man-powered aircraft has very promising potentiality as an educational project in aeronautical engineering ..."

He also sees it as a popular sport, while acknowledging the point made by Shenstone concerning the inhibiting performance and cost of machines which are considered excellent by today's standards.

A further review of the Kremer Competition rules may be carried out in the next year or so, and it is becoming increasingly probable that some relaxation of them will be implemented. Valuable data is becoming available concerning the desirability of two crew members, an argument which has yet to be satisfactorily resolved. Some form of assisted take-off could most appropriately be allowed should the Kremer prize not be claimed by 1978. It would be undesirable to ease the altitude restrictions, and any course changes which cut out the demonstration of manoeuvrability would tend to allow man-powered flight to become a simple test of pilot endurance without the added skill which the figure-of-eight demands.

There remains substantial scope for structural and aerodynamic improvements in man-powered aircraft;   indeed this exists in all branches of aviation.  A single seat aircraft designed as a contender for the Kremer prize by a group at the International Research and Development Company, Newcastle upon Tyne, and with which the author was associated, utilised a main spar in the form of a carbon fibre reinforced resin tube.  While carbon fibre in itself remains an expensive commodity, its price is likely to reduce substantially over the next few years and fibres will undoubtedly find application in model aircraft and home-builts, as well as in the less numerous man-powered types.

It is unlikely that the fixed wing man-powered aircraft will be supplanted in the foreseeable future.  Elliott's fluttered wing is the only untried new approach which may produce acceptable aerodynamic characteristics.  However, the possibility of improving the fixed wing machine aerodynamically to give a significant increase in efficiency is by no means ruled out.  One line which the author has briefly investigated is the design of a wing section for use close to the ground.  Man-powered aircraft use laminar flow aerofoils, in which energy dissipation is kept to a minimum by precise design and equally precise manufacture to meet the close design tolerances.  These aerofoils (sometimes known as isolated aerofoils) are originally designed to operate in what is known as infinite airspace, a condition similar to that of an aircraft flying at an altitude of hundreds or thousands of metres.  The operating characteristics of a wing flying very close to the ground are very different;  this is evident from the effect of ground proximity on induced drag, described earlier.

The tools have long been available with which an aerodynamicist can faithfully represent an aerofoil operating close to the ground, or 'in ground effect'.  What has until recently not been in existence is a theory with which the aerodynamicist could design such an aerofoil with a predicted optimum performance.  Indeed, the mathematical complexities of such a theory are too great to enable it to be applied at present to all but the most commercially attractive uses of aerofoils;  in the cases of the Company sponsoring the development, this happens to be cascades of turbine blades.

Experimental work on these blades has confirmed a theory in which one designs the displaced profile - a wing section plus its boundary layer of slow moving air close to the surface - from a number of prescribed boundary layers appropriate to the conditions under which the blade will operate.  One note-worthy feature of the theory is the representation of these boundary layers in a way which allows this complete control necessary when designing aerofoil sections of very low drag and corresponding high lift.

It will probably be several years before a theory along these lines is generally available for use in an application such as aerofoil design for man-powered aircraft.  However, such a theory, as well as offering the obvious aerodynamic advantages, may also allow one to take into account the strength requirement of the wing.  For example a fairly thick wing section with good strength characteristics in torsion may be desirable and one would be able to equate this to the corresponding aerodynamic properties required, and hence obtain the best structural/aerodynamic 'package'.

*  *  *

The last word is given to a futuristic account of lunar man-powered flight which recently appeared in a serious American magazine devoted to the assessment of scientific research and development.  Based on passages of a book entitled *Where the Wind Sleeps*, it describes a flight in one of the large pressurised domes erected so that man could live on the moon without cumbersome personal life support systems.  The sensation of low gravity flying is described vividly as follows:

"You are a bird, but not simply that.  Physically yes, though with a man's thoughts, instincts and emotions.  At first you feel like a bird and just a little silly because of those fabric wings strapped to your arms.  They are about fourteen inches wide and protrude several feet from the tips of your fingers.  The 'foot feathers', not feathers at all but rudder/ elevators made of the same neonylon that forms your wings, are stiffer than the wings since they are made for steering and provide very little lift. Then you jump a little and you *don't* come down.  You glide, hook a mild updraft - an (artificial) thermal or warm current - and begin to climb without even moving your arms.  Manoeuvre those wings to generate forward speed while keeping the lift and you really are a bird, for you are climbing under your own power, then soaring, dipping, wheeling, climbing again, diving a little, gently for the first time - and then you are laughing, ... laughing because man cannot fly but you are not a man any more.  You are the best of man and the best of bird.  Flying man."

*  *  *

# APPENDIX I

# Patents

Probably because of its obvious appeal to cranks and eccentrics, man-powered flight attracted more applications for patents in the nineteenth century than any other form of flight utilising steam, petrol, electric or other generation techniques. Around the turn of the century the application rate appears to have been particularly high - flight by any means was likely to succeed soon and muscle power was but one of a host of methods for achieving this.

The desire to protect ones inventions against exploitation by others was, of course, not the main motivating force behind this growth of applicants. Financial rewards for those who hit upon the correct combination of motor and machine for flight were undoubtedly principally responsible.

Patent applications in this field were considerably reduced once powered flight using the internal combustion engine had been shown to be successful and of much more practical use. A few unconventional layouts, and more usually details of some new gearing mechanism or a structural refinement, have made up the few patents reported in the past sixty years.

A complete record, including full descriptions of the inventions claimed concerning man-powered aircraft would stretch to many pages, but the following selection is representative of the quantities and, it is hoped, quality produced in the corresponding eras, a number of the more interesting samples being described in some detail. In sifting through the patents, it became evident to the author that they bore little or no resemblance, except in one or two cases, to the more promising lines of development which man-powered flight has taken in the past forty-five to fifty years. This is due to the fact that the technology has filtered through from powered and glider aircraft fields, and more often than not the patent specifying man-power as the source of propulsion originated from outside the aircraft industry or its associated institutions.

Patents and patent applications are listed in approximate chronological order, with the name of the patentee(s), patent number and date. The country of origin is given if the patent originated outside the United Kingdom.

* * *

Lawrence Holker Potts, Patent No. 9642, 1843.
"Balloons are employed for drawing cars on tension rails. A machine is described consisting of treadles, handles and cranks, enabling men to utilise all their strength in actuating wings and oars."

William Crofton Moat, Patent No. 9856, 1843.
"Machine, initially man-powered, made of an oblong framework of wood. In (accompanying) drawings four men are shown turning a crank, which rotates two frames on either side of the machine, each carrying four 'flappers'. These

315

give lift on the downstroke but rise edgewise to give lower resistance.
Undercarriage and rudder are provided."

William Edward Newton, Patent No. 11578, 1847.
"Use of small wings on the side of balloons, rotated to enable balloon to
maintain any desired altitude.  Reference is made to possible use of power
other than muscular."

Benjamin O'Neale Stratford, Earl of Aldborough, Patent No. 224, 1854.
(Improvements, No. 425 in 1855).
This inventor proposed a man-powered aircraft, either aided in flight by an
elongated balloon, or solely by wings.  The wings are intended to act in a
manner similar to birds, based on theories of bird flight put forward by the
inventor.  A tail is described, which may act as a rudder, striking downward
when the vessel is rising, thus compressing the air beneath, as the inventor
believed birds do, "especially pigeons".
"Any number of persons on board may aid propulsion, each having a separate
wheel".

Prosper Guilhaume Dartiguenave, Provisional Patent No. 1334, 1854.
Two objects similar to parachutes are placed one above the other, and
alternately pulled up and down.  These are powered, but wings are also provided
to turn the machine;  these are operated by the pilot.

Benjamin O'Neal Stratford, Earl of Aldborough,  Patent No. 2062, 1856.
Further improvements to Patent No. 224, 1854.  He appears now to suggest an
engine to provide propulsive power.  There are also improvements to the
passenger accommodation and other minor features.

Jean Baptiste Justin Lassie, Patent No. 2154, 1856.
Most original - Lassie proposed a cylindrical aerostat (balloon) having a
helical screw projecting from its outer surface.  A smaller cylinder is
provided within the balloon, in which the crew walk round in a similar manner
to "convicts in a treadmill".  This rotates the main balloon, which, by action
of the screw, advances through the air.  The balloon proposed was 257 metres
long, with a diameter of 28 metres.  Three hundred men would be employed, half
of these being on duty at any one time!

Viscount Carlingford, Patent No. 2993, 1856.
Fixed wing machine with screw propeller driven by hand cranks.  Optimistically,
he anticipated a range of 80 - 100 km, with take-off aided by a downhill
gradient.

Earl of Aldborough, Patent 1054, 1857.
(Further improvements on his earliest patent)
Crew weight is used for producing a rotary motion in the balloon, also to
operate wings and propellers.

Joseph Scott Phillips, Patent No. 2420, 1861.
Winged wheels rotated by hand (or steam) with feathering facility, similar to
paddle wheels.

George Davies, Patent No. 298, 1864.
"Machine with two wings and a tail, which are moved horizontally, vertically
and torsionally.  Whalebone strip and silk are used for wing covering.  The
operator uses hand wheels to move wings or tail."

Richard Brooman, Patent No. 2030, 1864.
"Extending arms are fixed at a distance from the body and terminate in wings
or flappers.  Springs are fitted at the connection of the wings and arms, and
the aeronaut works the flappers by means of chords."

Marc Antoine Francois Mennons, Patent No. 2299, 1864.
This patent describes a flapping wing aircraft actuated by the pilot.  The
machine is started by rolling down an incline.

Gustave Wilhelm Rothleb, Patent No. 1037, 1865.
"Strong, light frame surrounds body of person who is meant to fly and framing
is prolonged in front and behind about 5 ft (1.5 metres).  A windrose (having
arms like the sail of a windmill) is employed in front, and offers very little
resistance to the wind, as it is free to turn on its axis.  Revolving wheel
having feathered spokes acts as rudder.  Forward motion from wings actuated by
pilots' arms and spiral springs.  Two elastic reservoirs of hydrogen are used
to assist in giving lift."

James William Butler, Patent No. 1148, 1866.
'Manumotive Flying Machine'.
"Aeroplane is attached across (the pilots') shoulders.  Operator pulls wings
down, returned by springs.  Operator conveyed rapidly down a hill on a carriage
to gain momentum.  Start may also be obtained by jumping from a height, or by
being towed at high speed behind a rope."

Francis Herbert Wenham, Patent No. 1571, 1866.
"Two or more aeroplanes are arranged one above the other and support a
framework or car containing the motive power.  When the manual power is
employed, the body is placed horizontally, and oars or propellers are actuated
using the arms"  (Wenham also suggested using steam).

Thomas Craddock (Provisional) No. 1982, 1867.
"A small shaft is fixed to the person's body and rotated by means of clutch
pulleys, which are actuated by the feet and hands pulling on chords, with
springs aiding the return.  The rotation of this shaft actuates the wings,
producing an up and down, as well as a swivel, action, thus pulling the person
forward, as well as supporting him.  The wings are made like a lady's fan.
Steam may be used, enabling one to increase the number of wings."

Charles Green Spencer (Provisional refused) No. 1178, 1868.
"A T-shaped web is proposed having vertical and horizontal limbs which merge
into a cigar-shaped tube with a pointed nose.  The front portion houses the
pilot, who actuates two pairs of wings.  A rudder is provided."

Richard Harte, Patent No. 1469, 1870.
"A machine having wings and a two-bladed propeller, and a wheeled undercarriage.
The propeller will have variable pitch, and steam or man-power may be employed."

Philip Brannon, Patent No. 3974, 1877.
"Supporting surface resembling a house roof, or upturned boat, so as to act
both for propulsion and lift by its concave interior surface, in producing an
inward pressure on the adjacent air".  Propellers are described, adjustable in
various directions, driven by treadles or hand wheels.  Stored energy is
proposed for short journeys, but the storage method is not specified;  also,
steam or air engines are suggested as power sources.

William Jackson, Patent No. 513, 1878.
"Hand-turned propellers for balloon lift and forward motion".

Alfred Eugene Pichou, Patent 326476 (France), 1883.
Proposed a man-powered rotorcraft, named 'L'Auto Aerienne' which he tested
unsuccessfully in 1902.

William Cornelius, Patent No. 2589, 1884.
"Two wings are fulcrumed on the free ends of a U-shaped frame, a seat is
provided within the frame for the aeronaut, whilst he actuates the wings by
handles attached to the inner ends of the wing cord rod.  A hinged tail is
actuated by cords attached to his feet", *(see Fig. A1)*.

James Armour, Patent No. 14038, 1884.
"A pair of wings are hinged to a spring blade and actuated by man-power – each
pair of wings consisting of a main wing and a supplementary wing.  Steam may
also be used."

Arthur Charles Henderson, Patent No. 5118, 1885.
Suggested attaching wings to a frame below a balloon and working them by a hand
lever.

*Fig. A1*  William Cornelius' 'Flying Machine', 1884


    The second half of the 1880s saw a large number of applications of muscle
power to balloon propulsion.  Patents covering these are as follows:

    Edward Newall Molesworth-Hepworth.  Patent 7837, 1886.
    Sydney Lemmon.  Patent 13901, 1886.
    Wald Gustafson.  Patent 8255, 1887, *(see Fig. A2)*.
    Joaquim Ignacio Ribeiro.  Patent 8386, 1887
    George Seaborn Parkinson.  Patent 508, 1888
    William John Bastard.  Patent 3957, 1889.
    Herman and Albert Rieckert.  Patent 4811, 1889.
    Frederick William Zimer.  Patent 7427, 1889.
    Ernest Howard Grey.  Patent 5404, 1890.
    Steward Cairncross.  Patent 11455, 1890.

Of these ten projects, only Parkinson's errs from the standard techniques such
as hand cranking or pedalling.  He proposed propulsion by the reaction of three

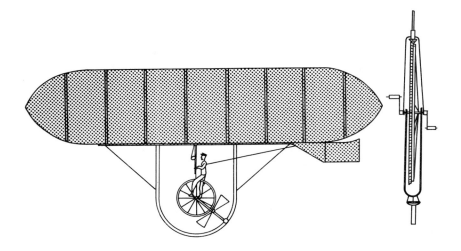

*Fig. A2*  The Wald Gustafson 'Navigable Balloon'.
Subject of Patent 8255 in 1887.

air jets, the air being forced through holes at the stern and side of the
balloon by "a fan or blower actuated by the pilot".

        The patents described below, filed later than those listed so far, are in
general, of greater interest.

        Patent No. 4567 was granted in March 1893 to H. Vaughan.  This covered an
apparatus for driving the propellers or paddle wheels of an aircraft (or,
states the patent, 'a boat').  It was based on the use of a gear wheel and
ratchet system, to which was connected a handle.  Reciprocal motion of the
handle resulted in continuous rotation of the gear wheels.  It was suggested
by Vaughan that each crew member could use two of these handles, in a similar
fashion to that used in rowing, to propel a machine, all the separate units
being connected to a single propeller shaft.

        Patent No. 4589, also dated March 1893, covered what was for that time
quite a progressive idea to utilise man power indirectly to propel a man-
powered aircraft.  This invention of J.O. O'Brien's was unfortunately designed
to flap the wings of an ornithopter rather than to drive a propeller, which
may have been more successful.  The wings of O'Brien's machine were actuated by
a compressed air motor supplied from a reservoir behind the pilot.  The
reservoir was replenished by air pumps actuated by hand and a treadle mechanism.
All of the machinery was supported on the back of the pilot, who was supposed
to run along the ground to start the mechanism.  An interesting feature (which
was really incidental to the operation of the main propulsive system) concerned
the wings of this machine.  Normally the upswing of a wing does not produce
lift, and any way in which the effort needed to return the wing to its upper
position can be minimised, must be to the pilot's advantage.  O'Brien conceived
that cloth valves within the wing surface could be used to assist this, and the

valves were designed such that on the downswing the complete surface of the
wing was sealed.

Later in 1893, W.P. Thompson related the use of paddle wheels as propulsion
units to man-powered flight driven by pedals. The wheels were driven
independently and were designed to reduce the pressure beneath the front of
the wing and increase it towards the rear. Patent No. 16269 was granted to
cover this invention.

In 1895 an Englishman named Moore was granted Patent No. 6 on January 1st
of that year[1]. Moore's machine embodied flapping wings, their design being
"in imitation of those of the flying fox". The machine could be man-powered,
but the purpose of the invention was concerned with the use of "an electro-
motor, using electro-magnets which in turn alternately attract and repel each
other". By means of this system cranks leading to the wings effected the
flapping mechanism.

Gough was granted Patent No. 459 in the same year, covering a device which
obviously owed its conception to the simple governor valve seen on steam
engines and the like. The mechanism involved is best likened to the rapid
opening and closing of an umbrella. Again the inventor refers to reciprocation
of the device by "hand or suitable power".

The bicycle was by now a comparatively cheap and easily mountable device,
closely resembling in layout and drive system  our contemporary machines.
Naturally these were, to many inventors, a Heaven-sent solution, or at least
aid with which to overcome gravity. Cycles were later (see Chapter 4) to
become the basis of a large collection of man-powered aircraft, and the patents
filed during 1895 include two which touch on, in varying degrees, the bicycle/
tricycle concept as a basis for the drive unit of an aircraft. Bodding, who
was granted Patent No. 15,019 for his variation on a pedal propulsion system,
only considered the aircraft as one of three types of transport to be
propelled by this means. A more specific invention is described in Patent No.
17,119, registered under the name of E.J. Pennington. The cylindrical object
attached to the rear axle of the bicycle *(Fig. A3)* is a motor of unspecified
type, and the actual function of the pilot is not specified in the description
of the patent. It is suspected that the so-called motor is purely a gear
device for transferring the axle rotation into motion, via the vertical shaft,
to drive the propeller. The machine sports a rudder at the front end,
controlled by the handlebars and two sets of wings. The patent description
concludes optimistically with the statement that "when the cycle is driven
rapidly forward by the motor and propeller, the aeroplanes raise the machine
and rider from the ground".

Several patented designs of man-powered aircraft appeared to be based on
an almost complete disregard of the comfort or convenience of the pilot, which
would have had some substantial effect on his maximum power output and
concentration. Two of these are particularly interesting.

*Fig. A4* shows a machine patented by C. Danilevsky in 1897 (Patent No. 24,532)
The use of a balloon to aid flight was not new, but Danilevsky's patent was

---

[1] *At this time the patent sequence of numbers ran for one year only, thus in*
*1896 for example, Patent Nos. 1 - 17,649 may be granted, and in 1897 patent*
*numbers could run from 1 - 23,468.*

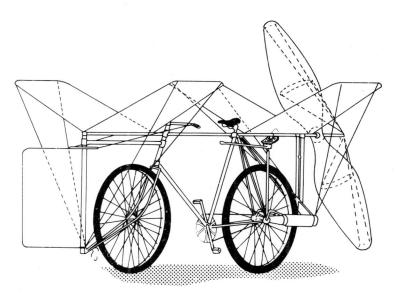

*Fig. A3*  E.J. Pennington's flying bicycle, invented in 1896.

concerned more with the system of levers whereby the pilot could manipulate the balloon incidence, and the use of arms and feet to flap the two wings. Specifically here, the balloon is not sufficiently buoyant to support a man without the assistance of some additional propulsion.  A seemingly unnecessary detail is the use of stilts by the pilot when running along the ground prior to take-off.

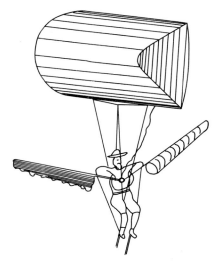

*Fig. A4*  C. Danilevsky's interpretation of balloon-assisted flight, in a patent dated 1897.

An even more difficult operating position, surely requiring a gymnast, was specified by Armitage in the patent (No. 20,739) granted to him in 1899. The description of his invention reads as follows:

"The device is made in imitation of the wings of a bird, and is used preferably without an aerostat (balloon) but may be used with one. Each wing comprises two ribs $a$, $b$,[1] respectively working on a shaft, $c$, and the extremities of the two ribs at one end of the shaft are connected to one stirrup, $e$, while those of the ribs at the other end of the shaft are connected to another. Similarly, the parts of the ribs beyond the fulcrum are connected to handles $n$, so that, by bringing the hands and feet together, the wings are brought into the position shown by the dotted lines to float and propel the machine, while by moving one hand or foot before or more vigorously than the other a certain amount of inclination is given to the wings to direct the machine upwards or downwards."

In the same year as Armitage, E. Edwards described in Patent No. 20,858, a screw propeller system which he claimed would improve the performance of both ships and man-powered aircraft, or more specifically, "water-velocipedes and airships". (The water velocipede was no doubt an early form of self powered paddle boat.) Drawings of the invention suggest that a number of bicycle frames mounted in tandem were slung beneath an airship. The wheels of the bicycles were replaced with one long Archimedes screw propeller, geared such that a contribution to the power for propulsion was made by each cyclist.

Unfortunately, bright new ideas were not lurking, ready to be announced at the beginning of the twentieth century. The year 1900 saw two ornithopter patents granted, No. 17,371 to Lehmann and No. 18,287 to Vergara.

Lehmann claimed to model the wings of a bird, the wings having imitation feathers at the trailing edges. They also sported a novel double skin, between which Lehmann claimed that a vacuum formed would assist in providing lift. The wings were intended to take up the shape shown in *Fig. A5*, the upper sketch showing the position adopted by the wing during the down-stroke, and the lower one representing air flow during the commencement of the up-stroke. In the

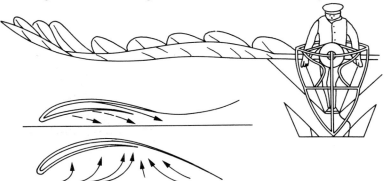

*Fig. A5*  E. Lehmann's ornithopter of 1900. A treadle mechanism was used to actuate the wings.

---

[1] *The letters refer to the drawing accompanying the patent. It is not reproduced here.*

sketch of the machine included with the patent, the pilot is shown to be
operating the wing flapping mechanism with both arms and legs.  The main spar
was moved upwards by depressing handles, and a treadle was used to bring the
wings back to their lowest position.

Vergara's design was based on the principle adopted by Gough in 1895,
detailed above, only instead of the complete umbrella type sail proposed by
Gough, a wing was to be attached to each side of the central shaft, and the
reciprocating motion was utilised to provide lift.  A feature whereby the wings
could be 'feathered' during flight was incorporated in Vergara's proposed
aircraft, although it is uncertain what function this facility served.  The
patent also stated that the wings could be supplemented by a screw propeller,
and steering was effected by a rudder.

M. Léger designed and patented what appears to have been the first man-
powered helicopter of recognisable layout, albeit without a stabilising rotor,
*(Fig. A6)*.  The machine derived its power via two contra-rotating propellers.
Apparently the propeller shaft could be tilted so as to provide limited
directional control.  The construction of the propellers was based on a steel

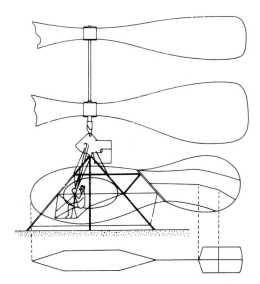

*Fig. A6*  M. Leger's helicopter, which utilised arm-power, 1901.

framework covered with aluminium sheet, and their rigidity was such that the
pitch could be manually adjusted by the pilot while the machine was in flight.
The cockpit had glazed openings and the undercarriage took the form of a tripod
with a hydraulic piston to deaden the landing shock.

Patent No. 18,768 granted to M. Bourcart in 1903 relates to a simple method
for the excitation of two propellers.  A double cranked shaft pedalled by the
pilot had a bevel gear at each end, driving two shafts inclined at about
15 degrees to the vertical.  At the top of these two shafts were mounted
propellers which were two-bladed and so synchronised as not to touch one
another during rotation.  Optional motor propulsion was catered for in

Bourcart's design.

A compact form of 'birdman' outfit was proposed by Brandl and detailed in Patent No. 9804 dated 1906. This arrangement consisted of wings fitted to the pilot's arm using a special frame, while his forearm was bent. The elbow fitted into a cup, and a small lever operated by the hand spread the wing, which was held in position using rods projecting laterally from the heels of the fliers' boots.

The continuing success of powered aircraft did not greatly deter inventors in the field of man-powered flight. The brothers U. and G. Antoni were granted a patent, No. 25,518 in 1907 for a design of ornithopter which was, for man-powered flight, supported beneath a balloon. (If a motor was used, the inventors considered that this would enable the machine to fly without the assistance of a 'lighter-than-air' device). The layout of the ornithopter resembled one of Leonardo da Vinci's concepts, the pilot using arms for flapping the wings and his feet for controlling the fan-tail type rudder. The designers had access to a much greater choice of materials and proven mechanical linkages than da Vinci, of course, and full use of these was made in attempting to optimise wing structure, with a rigid leading edge, and constructing an efficient lever system to manipulate the wings.

* * *

Up to this time patent offices in Europe could, to some extent, be regarded as clearing-houses for information. This arose primarily because many of the inventors had little or no hope of ever translating their ideas into hardware. In this case a patent afforded some protection and might in the future provide a lucrative source of income should an enterprising businessman wish to manufacture the patented device under licence.

Around 1910, as far as man-powered flight was concerned, this position considerably altered, and events such as the Peugeot Competition showed that the opportunity to attempt a flight was open to most people. The press and societies became the platforms whereby the layman could obtain assistance and moral support in his attempt to construct a machine, and patent cover, although remaining important, was secondary to the desire to perform a feat of flying. As a result, the history is readily traceable in places other than the Patent Office, and the number of patents being granted directly related to this topic has become negligible.

* * *

# APPENDIX II

# Conversion Factors

For several years, the engineering industry, and, more recently, companies selling directly on the domestic market, have been changing the system of measurement from 'Imperial' to 'Système Internationale (SI)'. This has led to displacement of the foot by the metre, the pound weight by kilograms, etc.

Numerical data in this book is presented using the SI system, and for those less familiar with this system, a number of old 'Imperial' measures and their SI equivalents are given below.

## SPACE AND TIME

| Length | 1 inch | = | 25.4 millimetres (mm) |
|---|---|---|---|
| | 1 foot | = | 0.305 metres (m) |
| | 1 yard | = | 0.914 metres |
| | 1 mile | = | 1.609 kilometres (km) |
| | | | |
| Area | 1 square foot | = | 0.093 square metres ($m^2$) |
| | 1 square inch | = | 6.452 square centimetres ($cm^2$) |
| | | | |
| Volume | 1 cubic foot | = | 0.028 cubic metres ($m^3$) |
| | | | |
| Velocity | 1 foot/second | = | 1.097 km/h |
| | 1 mile/hour | = | 1.609 km/h |

## MECHANICS

| Mass | 1 ounce | = | 28.35 gramme (g) |
|---|---|---|---|
| | 1 pound | = | 0.454 kilogramme (kg) |
| | 1 ton | = | 1.016 tonnes |
| | | | |
| Mass flow rate | 1 pound/second = | | 0.454 kg/s |
| | | | |
| Density | 1 pound/cubic inch | = | 27.68 g/$cm^3$ |
| | 1 pound/cubic foot | = | 16.02 kg/$m^3$ |
| | | | |
| Momentum | 1 foot-pound/second | = | 0.138 kg m/s |
| | | | |
| Force | 1 pound force | = | 4.448 Newtons (N) |
| | | | |
| Torque | 1 pound force-foot | = | 0.138 kgf-m |
| | | | |
| Pressure/ | 1 pound force/square inch | = | 0.07 kgf/$cm^2$ |
| Stress | 1 pound force/square foot | = | 4.882 kgf/$m^2$ |
| | 1 inch of water | = | 1.868 mm of mercury (mm Hg) |

Dynamic viscosity        1 pound force-second/square foot  =  47.88 Ns/m$^2$
                         (1 Ns/m$^2$  =  10 poise {P})

Kinematic viscosity      1 square foot/second      =    0.0929 m$^2$/s

Energy and work          1 foot-pound force        =    1.356 Joule (J)
                         1 kilowatt-hour           =    3.6 J

Power                    1 horse-power             =    0.745 kW.

* * *

# APPENDIX III

# Abbreviations

| | |
|---|---|
| AFIAS | Associate Fellow of the Institute of Aerospace Sciences (USA). |
| ARB | Air Registration Board (GB) — now the CAA (Civil Aviation Authority). |
| ASI | Air speed indicator. |
| AUW | All-up weight. |
| BAC | British Aircraft Corporation. |
| CAI | Canadian Aeronautical Institute (later to become CASI, Canadian Aeronautics and Space Institute). |
| FAA | Federal Aviation Administration. |
| FAI | Federation Aeronautique Internationale. |
| HPA | Hertfordshire Pedal Aeronauts. |
| HMPAC | Hatfield Man-Powered Aircraft Club. |
| IATA | International Air Transport Association. |
| I.Mech.E. | Institution of Mechanical Engineers (London). |
| LSARA | Low Speed Aerodynamics Research Association. |
| MAPAC | Man-Powered Aircraft Committee (Cranfield). |
| M.o.S. | Ministry of Supply (1957). |
| NACA | National Advisory Committee for Aeronautics (USA). |
| RAE | Royal Aircraft Establishment. |
| R.Ae.S. | Royal Aeronautical Society (London). |
| SUMPAC | Southampton University Man-Powered Aircraft (Club). |
| USAF | United States Air Force. |

* * *

# APPENDIX IV

# Bibliography

K. Buttenstedt. On the solution of the problem of flight. (In German) Der Stein der Weisen. Vol. 10. (No date given)

K. Steffen. A wind flying machine. (In German) Der Stein der Weisen. Vol. 23. (No date given)

A. Pénaud. Can man fly without an auxiliary motor? (In French) L'Aéronaute. Vol. 4, No. 6, p.39 - 45, June 1871.

Moore. Report on experiments to ascertain the power necessary for flight. Aeronautical Journal. Vol. 1, No. 4, p.2, 1897.

Marshall. Mechanical flight. Aeronautical Journal. Vol. 2, No. 7, p.51, 1898.

Moore. Experiments with flapping wings. Aeronautical Journal. Vol. 13, p.19, 1909.

P. Orlovskii. Is aerial flight solely by man's muscular exertion possible? (In Russian) Bibl. Vozdukhoplav. No. 1, p.44 - 46, 1909.

Winged aeroplanes by J. Bourcart, Colmar. (In German) Flugsport. Vol. 1, No. 24, 19th November, 1909.

H. La V. Twining. Can a man fly with wings? Aeronautics. Vol. 7, No. 1, p.2 - 4, July 1910; No. 2, p.39 - 41, August 1910; Vol. 8, No. 6, p. 189 - 190, June 1911.

Puiseux foot driven monoplane. (In German) Flugsport. Vol. 3, No. 2, 18th January, 1911.

The Peugeot prize of 10 000 francs. (In German) Flugsport. Vol. 4, No. 5, 28th February, 1912.

Flight by man power. Aero. Vol. 6, No. 108, p.63, March 1912.

Engineless flying machines and the first flight with a bicycle. (In German) Flugsport. Vol. 4, No. 12, 5th June, 1912.

Muscle as a motive power in flight. Scient. Amer. Vol. 106, No. 22, 1st June, 1912, New York.

New experiments by G. Poulain. (In French) L'Aéronautique, Paris. 26th October, 1919.

On the flying bicycle 'Aviette'. (In German) Flugsport No. 14, 7th July, 1920.

M. Flack.  The human machine in relation to flying.  Aeronautics.  Vol. 19,
No. 369, p.342 - 345, 11th November, 1920, London.  Aeron. Journal.  Vol. 24,
No. 120, p.650 - 663, December 1920, London.

G. Poulain.  Poulain's air bicycle.  Aeronautics.  Vol. 21, No. 404, p.30,
14th July, 1921.

A. Lippisch.  Theory of flapping flight.  Flugsport.  17th June, 1925.  (also
NACA Tech. Memo. 334).

O. Ursinus.  Wenn man ein Muskelflieg baut (If man builds a muscle powered
aircraft).  Flugsport.  Vol. 25, p.208 - 211, 1933.

R.E. Snodgrass.  Animal flight.  J. Roy. Aero. Soc.  Vol. 37, No. 266,
p.113 - 144.

R. Kronfield.  New possibilities of human flight.  (In German)  Flugsport.
No. 11, p.210, 24th May, 1933.

A. Lippisch.  The airskiff:  observations on human flight.  (In German)
Flugsport.  No. 15, p.311, 19th July, 1933.

Competition of the Polytechnical Society at Frankfurt am Main for a flight by
means of one's own muscle power.  (In German)  Flugsport.  No. 15, p.315,
19th July, 1933.

The Piskorsch rotating wing aeroplane.  (In German)  Flugsport.  No. 18, p.397,
30th August, 1933.

E. Everling.  Flight by human muscle power.  Luftwissen.  Vol. 1, No. 2,
p.35 - 37, 15th February, 1934.  Air Min. Trans. No. 233.

H. Haessler.  On the feasibility of muscle powered flight.  (In German)
Flugsport.  No. 1, p.2, 10th January, 1934.

O. Ursinus.  Thoughts on muscle power flight.  (In German)  Deutsch Luftwacht,
Luftwelt edition, Berlin.  Vol. 2, No. 4, 1935.

H. Gropp.  Aircraft propulsion by man power.  (In German)  Flugsport.  No. 3,
6th February, 1935.

H. Gropp.  Thoughts and figures on an aircraft driven by muscle power.  (In
German)  Flugsport.  No. 3, p.54, 6th February, 1935.

First controlled flight with human power.  (In German)  Flugsport.  No. 18,
p.395, 4th September, 1935.

R. Schulz.  First flights with human power.  (In German)  Deutsche Luftwacht.
Ausgabe, Luftwissen.  Vol. 2, No. 9, September 1935.

H. Gropp.  Muscular flight:  storing of energy for take-off.  (In German)
Flugsport.  Vol. 27, No. 24, p.562 - 564, 27th November, 1935.

H.G. Schulze, W. Stiasny.  Flug Durch Muskel Kraft.  (Muscle powered flight)
(In German)  Naturkunde und Technik.  Verlag Fritz Knapp, Frankfurt a.m., 1936.

W. Schmeidler. Possibility of human muscular flight. Air Min. Translation 298 from Luftfahrtforschung. Vol. 3, No. 1, p.12 - 13, January 1936.

Bibliography of articles and papers on muscular (flapping) flight from the Scientific and Technical press. Issued by the Directorate of Scientific Research and Technical Development, Air Ministry (Prepared by R.T.P.) No. 10, June 1936.

H. Seehase. Human powered flight: a constructive contribution. Flugsport. No. 18, p.491, 1937.

O. Ursinus. Versuche mit Energie-speichen (Research on energy storage). Flugsport. p.33 - 40, 1937.

O. Ursinus. Reports of the Muscular Powered Flight Institute, 1935 - 1937. (In German) Muskelfluginstitut (Polytechnische Ges., Frankfurt). 1937.

C. Silva. Gli studi sul volo muscolare. (In Italian) (The study of muscle powered flight in Italy) Mitteilungsblatt No. 7 der Internationalen Studien Kommission für den Motorlosen Flug. p.58, 1938.

D.R. Wilkie. The relation between force and velocity in human muscle. J. Physiol. Vol. 110, p.249 - 280, 1950.

I.N. Vinogradov. The aerodynamics of soaring bird flight. Aerodisamike Prits, Paritelie. 1951. (RAE Library Translation No. 846, 1960).

A.W. Raspet. Human muscle powered flight. Soaring. Vol. 16, p.18 - 19, 1952.

A. Raspet. Unsolved: the problem of Leonardo da Vinci, human muscle powered flight. Presented to Mississippi Academy of Science on 2nd May, 1952.

Liberatore. Special types of rotary wing aircraft. PB 111633 U.S. Department of Commerce Office of Technical Services, 1954. (Includes illustrations and details of various muscle powered aircraft).

B.S. Shenstone. Unusual aerodynamics. CAI Log. p.19, December 1954. (Review of the unresolved problems of low speed natural and pseudo-natural flight).

T. Nonweiler. The man-powered aircraft; a preliminary assessment. College of Aeronautics. Note No. 45, 1956.

R.G. Bannister. Muscular effort. Brit. Med. Bull. Vol. 12, p.222 - 225, 1956.

B.S. Shenstone. The problem of the very light weight highly-efficient aeroplane. Canadian Aeronautical Journal. p.83 - 90, March 1956.

T.R.F. Nonweiler. The work production of man: studies on racing cyclists. J. Physiol. Vol. 141, p.8, 1958.

T.R.F. Nonweiler. Man powered aircraft - design study. Roy. Aero. Soc. J. Vol. 62, No. 574, p722 - 34, October 1958.

D.R. Wilkie. The work output of animals: flight by birds and by man power. Nature. Vol. 183, p.1515 - 1516, 1959.

E.R. Kendall.  Is man powered rotating wing flight a future possibility?
J. Helicopt. Assoc. G.B. Vol. 13, No. 2, p.100 - 130, April 1959.

B.S. Shenstone.  Man powered aircraft.  Shell Aviation News.  No. 750, p.14 - 18,
April 1959

C.H. Naylor.  Is man powered rotating wing flight a future possibility?
J. Helicopt. Assoc. G.B.  Vol. 13, No. 5, p.294 - 304, October 1959.

B.S. Shenstone, R.H. Whitby.  Man powered helicopters.  J. Helicopt. Assoc. G.B.
Vol.13, No. 5, p.305 - 315, October 1959.

T.R.F. Nonweiler.  The fixed wing approach to man powered flight.  The New
Scientist.  Vol. 6, No. 162, p.1291 - 1293, December 1959.

A.M. Lippisch.  Man powered flight in 1929.  J. Roy. Aero. Soc.  Vol. 64,
No. 595, p.395 - 398, July 1960.

B.S. Shenstone.  Man powered flight achieved in 1936?  The Aeroplane and
Astronautics.  p.228 - 229, 19th August, 1960.

B.S. Shenstone.  Engineering aspects in man powered flight.  J. Roy. Aero. Soc.
Vol. 64, No. 596, p.471 - 477, August, 1960.

D.R. Wilkie.  Man as an Aero Engine.  J. Roy. Aero. Soc.  Vol. 64, No. 596,
p.477 - 481, August 1960.

E. Bossi.  A man has flown by his own power in 1937.  Canadian Aeronautical
Journal.  Vol. 6, No. 10, p.395 - 399, December 1960.

F. Villinger.  Man powered:  fact and fancy.  The aeroplane.  p.378 - 9,
6th April, 1961.

A.T.E. Bray and A.E.G. Strain.  List of references on man-powered flight and
the Kremer Competition.  Royal Aircraft Estab. Library Bibliography.  No. 228,
RAE Farnborough, April 1961.

R.H. Wickens.  Aspects of efficient propeller selection with particular
reference to man powered aircraft.  Canadian Aeronautical Journal.  p.319 - 330,
November 1961.

R. Graves.  Problems of a man-powered rotorcraft.  J. Roy. Aero. Soc.  Vol. 66,
p.707 - 712, November 1962.

J.J. Cornish and W.G. Wells.  Boundary layer control applications to man-
powered flight.  Canadian Aeronautics and Space Journal.  p.55 - 61, February
1963.

Man-powered flight:  State of the art.  Flight International.  p.325 - 327,
27th February, 1964.

C.H. Gibbs-Smith.  A brief history of flying:  from myth to space travel.
London.  H.M.S.O.  1967.

C.H. Gibbs-Smith.  Leonardo da Vinci's Aeronautics.  London.  H.M.S.O.  1967.

W. Czerwinski.  Structural trends in the development of man-powered aircraft.
J. Roy. Aero. Soc.  Vol. 71, No. 673, January, 1967.

C.H. Gibbs-Smith.  Sir George Cayley (1773 - 1857).  London H.M.S.O.  1968.

S. Richmond.  Man-powered flight - A new approach to an old dream.  J. Roy.
Aero. Soc.  Vol. 73, No. 703, p.615 - 619, July 1969.

W. Czerwinski.  Man-powered flight, its purpose and future.  AIAA Paper No.
70-879.  CASI/AIAA Meeting on the prospects for improvement in efficiency of
flight.  Toronto, Canada.  9th - 10th July, 1970.

K. Sherwin.  Man-powered flight.  Hemel Hempstead.  M.A.P.  1971.

T.R.F. Nonweiler.  A new series of low drag aerofoils.  Ministry of Aviation
Supply, Aeronautical Research Council, R and M No. 3618,  H.M.S.O. London, 1971.

P. Jarrett.  Pilcher and the multiplane - A neglected aspect of a pioneer's work.
Aero. J.  Vol. 76, No. 741, September 1972.

Announcement of £50 000 Kremer Competition.  Aero. J.  Vol. 77, No. 749, May
1973.

J. Potter.  Man-powered flight - the Jupiter project.  Aero. J.  Vol. 77,
No. 751, July 1973.

Anon.  Man-powered flight.  Roy. Aero. Soc., London.  1973.

H. Upenieks.  Man-powered flight.  The oscillating wing machine.  Aero. J.
Vol. 77, No. 754, October 1973.

Ann Welch.  Development of the competition glider.  Aero. J.  Vol. 77, No. 756,
December 1973.

J.S. Sproule.  The simple flying virtues.  Flight International.  Vol. 105,
21st March, 1974.

C.H. Gibbs-Smith.  Sir George Cayley, Father of aerial navigation (1773 - 1857)
Aero. J.  Vol. 78, No. 760, April 1974.

D.J. Betteridge and R.D. Archer.  A study of the mechanics of flapping flight.
Aeronautical Quarterly.  Vol. XXV, Part 2, May 1974.

Anon.  Proc. Man-Powered Aircraft Group Symposium.  Roy. Aero. Soc. London,
10th February, 1975 (7 papers).

J.C. Wimpenny.  Structural design considerations of man-powered aircraft.  The
Aeronautical Journal.  Vol. 79, No. 773, May 1975.

J. Potter.  Handling techniques for man-powered aircraft.  The Aeronautical
Journal.  Vol. 79, No. 773, May 1975.

K. Sherwin.  To fly like a bird.  Bailey Bros. & Swinfen Ltd., 1976.

* * *

# APPENDIX V

# Kremer Competition Rules

The following pages contain reproductions of the rules of the current main Kremer Competitions organised and administered by the Royal Aeronautical Society in London.

The '£50 000 Kremer Competition' which is fully international, requires completion of a figure-of-eight course. The rules are very similar to those published in February 1960, when the competition was first announced. However at this time the prize was only £5000 and entry was restricted to Commonwealth subjects.

The '£5000 Kremer Competition', involving completion of a 'slalom' course, was introduced in 1967. The prize money will be divided and given to the first three successful entrants. In this case the competition is restricted to United Kingdom and British Commonwealth subjects.

In addition to the two competitions detailed here, a '£1000 Kremer Competition' was initiated in 1976, to be awarded to the first United Kingdom entrant to make a flight having a duration of at least 3 minutes. The course may be selected by the entrant, with the proviso that the aircraft exceeds an altitude of 2 metres at some point between take-off and landing.

An annual prize of £250 is also being offered by the Royal Aeronautical Society for the best project study by students at universities, colleges and polytechnics.

## THE £50 000 KREMER COMPETITION
### REGULATIONS

1. GENERAL

   The prize will be awarded to the entrant who first fulfils the conditions.

2. PRIZE

   The prize is £50 000 sterling.

3. ELIGIBILITY

   The competition is international and is open to individuals or teams from any part of the world. Rights of appeal will be governed by the Competition Rules of the AOPA and the Sporting Code of the Fédération Aéronautique Internationale.

4. CONDITIONS OF ENTRY

4.1 Aircraft

   4.1.1 The machine shall be a heavier-than-air machine.

4.1.2   The use of lighter-than-air gases shall be prohibited.

4.1.3   The machine shall be powered and controlled by the crew of the machine over the entire flight.

4.1.4   No devices for storing energy either for take-off or for use in flight shall be permitted.

4.1.5   No part of the machine shall be jettisoned during any part of the flight including take-off.

## 4.2  Crew

4.2.1   The crew shall be those persons in the machine during take-off and there shall be no limit set to their number.

4.2.2   No member of the crew shall be permitted to leave the aircraft at any time during take-off or flight.

4.2.3   One handler or ground crew shall be permitted to assist in stabilising the machine during take-off, but in such a manner that he is unable to assist in accelerating the machine.

## 4.3  Ground Conditions

4.3.1   All attempts, which shall include the take-off run, shall be made over approximately level ground (i.e. with a slope not exceeding 1 in 200 in any direction), and on a course to be approved by the AOPA or in conjunction with its authorised representatives.

4.3.2   All attempts shall be made in still air, which shall be defined as a wind not exceeding a mean speed of approximately 10 knots, over the period of the flight.

## 4.4  Course

4.4.1   The course shall be a figure of eight, embracing two turning points, which shall be not less than half a mile apart.

4.4.2   The machine shall be flown clear of and outside each turning point.

4.4.3   The starting line, which shall also be the finishing line, shall be between the turning points and shall be approximately at right angles to the line joining the turning points.

4.4.4   The height, defined as ground clearance, both at the start and the finish, shall be not less than ten feet above the ground; otherwise there shall be no restriction in height.

4.4.5   The machine shall be in continuous flight over the entire course.

## 4.5  Observation

Every attempt shall be observed by the AOPA or by any body or persons authorised by them to act as observers.  It may take place in the Competitor's own country if it is affiliated to the FAI.  In a country not so it could be advantageous to fly the course in a neighbouring country which is so affiliated.

## 5.  APPLICATIONS FOR ENTRY

5.1   Entry forms shall be obtained from, and returned to The Secretary, Man Powered Aircraft Group.  The Royal Aeronautical Society, 4 Hamilton Place, London W1V OBQ.

5.2 The entry fee shall be £1 (made payable to the Royal Aeronautical Society), which shall be refunded upon the attempt taking place.

5.3 Each entry form shall contain an application for Official Observation of the competitor's attempt.

5.4 The entrant shall undertake to abide by the conditions for Official Observation as set out on the entry form and shall undertake to defray all expenses incurred in connection with the Official Observation of the attempt.

5.5 Final notice of the proposed time and place of the attempt requiring Official Observation may, if so wished, be sent to the R.Ae.S. later than the Entry Form. It must in all cases be received at least thirty days before the proposed date for the attempt. This time is required by the AOPA to arrange for Official Observation. Applications will be considered in order of receipt.

5.6 The Entry Form or the final notice of the attempt must be accompanied by the sum of £15, made payable to the Aircraft Owners' and Pilots' Association.

5.7 Competitor's Annual Licence

This licence is required for all pilots taking part in the Kremer Competitions. It is not required for other flights. Application forms may be obtained as in paragraph 5.1.

## 6. GENERAL CONDITIONS

### 6.1 Insurance

The entrant must take out on behalf of himself, his pilot(s), crew, representatives or employees, an adequate insurance to indemnify the Society against any claims. Evidence that such insurance has been effected must be produced to the Official Observers before the attempt.

### 6.2 Eligibility

In any question regarding the acceptance of entries, eligibility of entrant, pilot, crew or aircraft under these Regulations, the decision of the R.Ae.S. shall be final.

### 6.3 Supplementary Regulations

The R.Ae.S. reserves the right to add to, amend or omit any of these regulations and to issue Supplementary Regulations.

### 6.4 Interpretation of Regulations

The interpretation of these Regulations or any of the Regulations hereafter issued shall rest entirely with the R.Ae.S. The entrant shall be solely responsible to the Official Observers for due observance of these Regulations and shall be the person with whom the Official Observers will deal in respect thereof, or any other question arising out of this Competition.

### 6.5 Revision of Regulations

These Regulations shall remain in force until such time as the Royal Aeronautical Society considers it necessary to amend them, or the prize has been won.

## THE £5000 KREMER COMPETITION
## REGULATIONS

### 1. GENERAL

The prizes will be awarded to the first three entrants who fulfil the conditions.

### 2. PRIZES

The prizes are:    First £2500;        Second £1500;        Third £1000.

### 3. ELIGIBILITY

The entrant, designer and pilot must be citizens of the United Kingdom or the British Commonwealth.  The aircraft must be designed, built and flown within the British Commonwealth.  Rights of appeal will be governed by the Competition Rules of the AOPA and the Sporting Code of the Fédération Aéronautique Internationale (FAI).

### 4. CONDITIONS OF ENTRY
### 4.1 Aircraft

4.1.1  The machine shall be a heavier-than-air machine.

4.1.2  The use of lighter-than-air gases shall be prohibited.

4.1.3  The machine shall be powered and controlled by the same crew over both parts of the flight.

4.1.4  No devices for storing energy either for take-off or for use in flight shall be permitted.

4.1.5  No part of the machine shall be jettisoned during any part of the flights including take-off.

### 4.2 Crew

4.2.1  The crew shall be those persons in the machine during take-off and flight, and there shall be no limit set to their number.

4.2.2  No member of the crew shall be permitted to leave the aircraft during take-off or while it is in flight.

4.2.3  One handler or ground crew shall be permitted to assist in stabilising the machine during each take-off, but in such a manner that he is unable to assist in accelerating the machine. The machine may be manually turned on the ground by the ground crew before the return flight.

### 4.3 Ground Conditions

4.3.1  All attempts, which shall include the take-off run, shall be made over approximately level ground (i.e. with a slope not exceeding 1 in 200 in any direction), and on a course to be approved by the AOPA or its authorised representatives.

4.3.2  All attempts shall be made in still air, which shall be defined as a wind not exceeding a mean speed of approximately 10 knots, over the period of the flight.

## 4.4  Course

4.4.1   The course shall consist of two flights in opposite directions, each including three turns made around three markers spaced at intervals of a quarter of a mile in a straight line. The starting point shall be on the extension of the line of the markers and may be as near to the first marker as the competitor wishes.

4.4.2   The course must be completed in both directions within an elapsed period of one hour from the start. The finishing line in each case shall be an extension of the line of the markers.

4.4.3   The machine shall be flown clear of and outside each marker, with the turn round the central marker to port in one direction and to starboard in the other.

4.4.4   The height of the machine above the ground shall be not less than 10 ft when passing the first and third markers; otherwise there shall be no restriction as to height, but after each take-off the machine must remain in continuous flight until the finishing line at the opposite end of the course is crossed.

## 4.5  Observation

Every attempt shall be observed by the AOPA or by any body or persons authorised by them to act as observers. It may take place in the Competitor's own country if it is affiliated to the FAI. In a country not so it could be advantageous to fly the course in a neighbouring country which is so affiliated.

Sections 5 and 6, relating to Application for Entry and General Conditions, are as for the £50 000 Kremer Competition.

\* \* \*

# APPENDIX VI

# Definitions

*(see also British Standard B.S. 185, Pt. 1:
Glossary of Aeronautical Terms).*

Aerostat:
: A balloon - that part which contains the lighter-than-air gas.

Ailerons:
: Movable surfaces used to control roll, and generally located at the wing trailing edge.

All-up Weight:
: The aircraft gross weight. For a man-powered aircraft this is generally the sum of the structural and pilot weights.

Aspect Ratio:
: The ratio of the span to chord of an aerofoil (wing). Man-powered aircraft are now characterised by a very high aspect ratio. 'Geometric' aspect ratio denotes the value obtained by physical measurement. 'Effective' aspect ratio takes into account changes in drag brought about by flying near the ground. These changes may be likened to the effect of an increase in aspect ratio.

Boundary Layer:
: The thin layer of air adjacent to all parts of an aircraft surface, in which viscous forces influence the motion of the the air.

Canard:
: An aircraft having its elevator in front of the main wings.

Chord:
: The distance between the leading and trailing edge of an aerofoil (wing).

Dihedral:
: The upward inclination of an aircraft's wings from their roots.
'Anhedral' is the corresponding downward inclination. Dihedral aids roll stability.

Drag:
: The component of any fluid - dynamic force, the direction of which coincides with that of the undisturbed flow (against an obstacle). (See Fig. A7).

Flutter:
: A sustained oscillation due to the interaction between aerodynamic forces, elastic reactions, and inertia.

Ground Effect:
: The proximity of the ground can alter the downwash angle behind a wing, reducing the component of the lift force in the direction of flow (induced drag). This phenomenon is very important to man-powered aircraft.

Incidence:
: The angle the wing chord-line makes with the direction of motion, (relative to undisturbed air).

Induced Drag:
: Drag due to lift, being the component of the lift force resolved in the direction of undisturbed flow.

Induced Drag Factor:
: A correction factor (normally between 1 and 1.1) applied to the expression for induced drag, to take into account differences in distribution of lift on a wing.

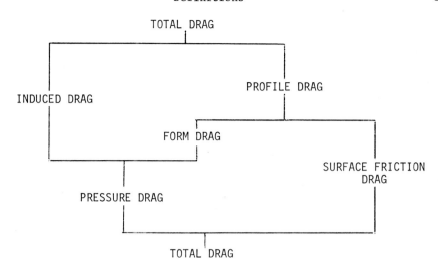

*Fig. A7* Breakdown of drag types acting on an aircraft,
(See definitions).

| | |
|---|---|
| Lift: | The vertical component of the total aerodynamic forces acting on an aerofoil or complete aircraft. |
| Load Factor: | The ratio of a specified external load applied to an aircraft, to the weight of the aircraft itself.  The external load may arise from gravity, aerodynamic forces, landing shocks, or any combination of these.  Limiting loads are described in a 'Flight Envelope'. |
| Magnus Effect: | The name given to the phenomenon by which a cylinder, rotating about its spanwise axis in an air flow, sets up a circulation and generates lift. |
| Ornithopter: | An aircraft lifted and propelled by flapping wings or other similarly moving attachments. |
| Pitch: | The motion of an aircraft about its lateral axis. |
| Pressure Drag: | The part of the drag due to the resolved components of the pressures normal to the surface. |
| Reynolds Number: | The ratio between viscous and dynamic forces on a body. |
| Roll: | Angular motion of an aircraft about its longitudinal axis. |
| Spoiler: | Hinged plates, normally in a position flush with the surface of a wing, and located near its trailing edge, which are raised to effect roll control.  They may replace ailerons, and can double as air brakes. |
| Stagger: | The amount one wing of a biplane is displaced forwards or rearwards, relative to the other wing. |
| Stall: | Aircraft behaviour brought about by loss of lift owing to the wing incidence being too large.  Loss of speed is |

                     followed by a lessening ability to control the aircraft.

Taper Ratio:            The ratio of the root chord to the tip chord.

Thickness/Chord Ratio:  The ratio of the maximum dimension of an aerofoil, measured perpendicular to the chord line, to the chord length.

Wing Loading:          The gross weight of an aircraft divided by its wing area.

Wing Section:          The cross-sectional shape of a wing, by which it is identified as being unique.  This shape largely determines the wing performance.

Yaw:                    Angular motion of an aircraft about its normal axis.

* * *

**ADDENDUM**

# Has the Kremer Competition Been Won?

On Tuesday 23rd August, 1977, Bryan Allen, piloting a canard monoplane, illustrated overleaf, completed a figure-of-eight course above Shafter Airfield in California, flying a distance of approximately 2.25 km. Carried out in the presence of official observers, the attempt on the Kremer prize has yet to be ratified by the Royal Aeronautical Society. However, regardless of the outcome, the flight has shown that, with what appears to be a comparatively unsophisticated design, controlled man-powered flight over a reasonable distance is possible.

The 'Gossamer Condor' was designed by Paul MacCready and Vern Oldershaw, MacCready being an aerodynamicist and former gliding champion. The aircraft has a wing span of 29.5 metres and an empty weight of approximately 32 kg (the pilot, an amateur racing cyclist, weighing 64 kg). Considerable use was made of thin-walled aluminium tubing in the structure, and the 3.65 metre diameter propeller was constructed using balsa wood.

Other data on the aircraft, which cost in the region of £14 000 to construct, is not yet available. However, one must now question the effect this will have upon man-powered flight. Major projects currently under way may be abandonned. Will there now be a rebirth of interest in man-powered flight as a sport? Can another goal, with even richer rewards for the competitors, be established?

If confirmed, Bryan Allen's flight has meant the end of one era for man-powered flight — what will the next bring?

25 August 1977

341

Bryan Allen piloting the 'Gossamer Condor'

# Index

Bernier, 19
Bernhard, R., 68
Besnier, 19
Betteridge, D.S., 236
Bibliography, 328 et seq.
BICH-18, 113
Bicycle, 65, 255, 320
Biplanes: 109
   Cheranovski, 113
   Fischer, 237
   Helbock, 84
   Leslie Smith, 258
   MIT, 295
   Parkes, 59
   Research, 272 et seq.
   Smolkowski, 219
Bird flight: 115
   Fullerton, 45
   Günter & Guerra, 131 et seq.
   Jordanglou, 135
   Magnan, 135
   Russian experiments, 56, 115
'Bird I', 86
Birds: 41, 115, 132
   condor, 41
   humming-bird, 133
   power output, 18
   wing loading, 45
   wing movement, 34, 41, 56 et seq.,
   115, 280
Bishop Wilkins, 18
Blades (propeller), 268 et seq.
   number, 142
   speed, 268
Bladud, 5
Blanchard, J-P., 23
Bolori, 19
Bonham, P., 233
Bonomi, V., 109
Borelli, 16
Bossi, E., 108 et seq., 217
Boundary layer:
   control of, 128, 136, 269
   laminar, 129
   on propellers, 269
   suction, 130, 151, 269
   transition, 184
Bourcart, J.J., 35
Bourcart, M., 323
Brannon, P., 317
Breant, 35
Breary, W.F., 36
Bredetzky, S., 21
Bristol Polytechnic, 279
British Aircraft Corporation, 155,
   285

Brooman, R., 317
Brown, D.S., 38
Brown, J.R., 138
Brunt, B., 24
Brustmann, M., 87 et seq.
Bryant, R., 258
Butler, J.W., 317
Buttenstedt, C., 41, 43


Cairncross, S., 318
Calgary, 219
Caliendi, S.C., 184
Canada:
   CASI, 12, 216
   Calgary, 219
   Czerwinski, 218 et seq.
   Expo '67, 220
   Ottawa Group, 221 et seq.
   research, 268
   Smolkowski, A., 219
   Wickens, R.H., 268
Canadian Aeronautics & Space Institute,
   12, 216 et seq.
Canadian Aviation Electronics, 220
Canard surfaces: 202
   biplane, 295
Canon Ogar, 20
Cardington, 176
Carnevali, U., 237
CASI plaque, 220
Casco, E., 111
Catapult launch, 111, 151
Cayley, Sir George: 32 et seq.
   helicopters, 33
   ornithopters, 32
Celebi, H., 19
Celio, H., 233
'Cellovel', 238
Cesari, R., 238
Chang Hêng, 4
Chang Yin, 4
Chanute gliders, 45
Chains:
   efficiency, 269 et seq.
Cheranovski, V.I., 113
Child, I., 258
Child-power: 33
   aircraft design, 266 et seq.
Chile University, 132
China, 2 et seq.
Chirambo, S., 232
Church, C., 186
Cierva Essay Competition, 250
Claudel, 35

— ∞ ✳ ∞ —